DATELINE CANADA
Understanding Economics through Press Reports

DATELINE CANADA
Understanding Economics through Press Reports

Peter Kennedy
Simon Fraser University

Gary Dorosh
Douglas College

PRENTICE-HALL OF CANADA, LTD. SCARBOROUGH, ONTARIO

To DATE
Who wishes she could understand, and

To DAD
Who always knew I could do it.

Canadian Cataloguing in Publication Data

Kennedy, Peter, 1943-
    Dateline Canada
ISBN 0-13-196774-6

1. Economics. 2. Canada—Economic conditions—
1965-   * 3. Canadian newspapers (English)—
Sections, columns, etc.—Finance.* I. Dorosh,
Gary, 1942-   II. Title.
HB171.5.K45   330.9'71'0644   C77-001679-0

© 1978 by Prentice-Hall of Canada, Ltd.
All rights reserved. No part of this book
may be reproduced in any form without permission
in writing from the publishers.

Prentice-Hall, Inc., Englewood Cliffs, *New Jersey*
Prentice-Hall International, Inc., *London*
Prentice-Hall of Australia, Pty., Ltd., *Sydney*
Prentice-Hall of India Pvt., Ltd., *New Delhi*
Prentice-Hall of Japan, Inc., *Tokyo*
Prentice-Hall of Southeast Asia (PTE.) Ltd., *Singapore*

ISBN 0-13-196774-6

1 2 3 4 5 82 81 80 79 78

Printed and bound in Canada by Webcom

Contents

Preface ix

Part I MICROECONOMICS

I-1 Supply and Demand 2
- a) Lumber 3
- b) Coffee 4
- c) Meat 5

I-2 More Supply and Demand 6
- a) The Futures Market 7
- b) The Exchange Rate 8
- c) The Money Market 9

I-3 Elasticity 12
- a) Public Transit 13
- b) Ferry Use 14

I-4 The Corn-Hog Cycle 15
- a) Pork Rise Chops Prices 16

I-5 Costs 17
- a) Growing Cherries 18
- b) Copper Smelters 19

I-6 At the Margin 21
- a) Air Cargo Operations 22
- b) Electric Metering 23

I-7 Taxes and Subsidies 25
- a) Non-Refillables 27
- b) Surplus Milk 28

I-8 Government-Fixed Prices 31
- a) Minimum Wages 32
- b) Rent Controls 33
- c) The Oil Market 35

I-9 Marketing Boards 37
- a) Fresh Eggs 38
- b) Food Scraps 40

I-10 Foreign Competition 42
- a) Double-Knit Bankruptcies 43

I-11 Imperfect Competition 45
    a) Blue Jeans Blues 47
    b) Car Glut 48
    c) Cash Rebates 50

I-12 Regulated Monopolies 51
    a) Railroads 52
    b) Hydro 54

I-13 Taxing a Monopoly 57
    a) Freeway Gas Stations 57

I-14 Price Discrimination 59
    a) First-Class Travel 60
    b) Excursion Rates 61

Part II MACROECONOMICS

II-1 Measuring Unemployment 64
    a) Dropouts Welcome 65
    b) Who are the Unemployed? 66

II-2 Unemployment Insurance and the Work Ethic 68
    a) The Work Ethic 69
    b) Unemployment Insurance 70

II-3 Consumer Behavior 71
    a) Consumer Behavior 72

II-4 The Inventory Cycle 73
    a) The Inventory Cycle 74

II-5 Fiscal Policy 75
    a) Balancing the Budget 76
    b) Tax Cuts? 77
    c) The Impact of Corporate Tax Cuts 78

II-6 Monetary Policy 82
    a) The Money Supply 84
    b) A Rise in the Bank Rate 86

II-7 Independence of the Monetary Authorities 88
    a) The Bank's Responsibilities 89

II-8 Demand-Shift Inflation 91
    a) Service Sector Growth 92

II-9 Stagflation 94
    a) Phillips Curve Makes Front Page 96
    b) Textbook Theory 98

II-10 Wage and Price Controls 99
    a) A Continuing Debate 100
    b) Old Policies Rejuvenated 103

II-11 Interest Rates 104
    a) Inflation and Interest Rates 106
    b) International Influences on our Interest Rate 107
    c) The Money Supply and Interest Rates 108
    d) A Proposal to Curb Inflation 109

II-12 Government Financing 111
    a) The Budget and Financing 112
    b) How to Raise $1/2 Billion 113
    c) Financing the U.S. Government 115

II-13 International Influences 117
    a) Pressures on Exchange Rates 119
    b) A Stronger Canadian Dollar 120
    c) A Too-Strong Canadian Dollar 121
    d) Should the C$ Be Devalued? 123

II-14 Fixed or Floating Rates? 126
    a) Floating Rate Suits Canada 127
    b) An Exchange Fight 129

II-15 Growth 131
    a) A Capital Shortage 132
    b) A Conserver Society 133

II-16 The Future: Theory and Policy 136
    a) Corporatism 136
    b) Corporatism Won't Work 139
    c) Separatists' Economics 140

Glossary 143

Acknowledgments

No authors are without debts. Thanks are due to Percy Christan-Quao, Rod Midgley, and Rocky Mirza who helped with classroom testing at Douglas College. Doug Beck, Don Gordon, John Herzog, and Steve Spector at Simon Fraser University, and Don Daly of York University offered both advice and encouragement. Howie Day and Don McRae at Vancouver City College generously made available to us their newspaper files. The greatest debt, however, is owed to those students at Douglas College who bore the burden of classroom testing; they uncovered many shortcomings. The errors they did not find, however, remain our responsibility.

Preface

Much of the information which people receive concerning economic decision-making in our society comes through the news media, especially newspapers; hence one of the most valuable skills a student of economics can develop is the ability to interpret newspaper articles relating to economics, root out assumptions contained therein, discard erroneous economic reasoning, evaluate the article's conclusions, and place those conclusions in their proper perspective.

To help students develop an ability to apply their knowledge of economics to newspaper articles, *Dateline Canada* presents typical news reports dealing with economic phenomena, then asks questions relating to the reports. These questions test students' interpretation of the contents of the articles, their perception of the assumptions embodied within them, and their understanding of the economic reasoning. The articles were carefully chosen to be suitable for this purpose; the questions associated with each were designed to draw out of that article as much of the economic terminology, reasoning, and controversy as possible, without depriving the student of the challenge associated with the interpretation and evaluation of the article.

The book is divided into two parts, one covering microeconomic topics, and the other covering macroeconomic topics. Within each half, the material is presented roughly in order of its normal appearance in a principles textbook. It must be stressed that the topic coverage is not fully comprehensive. Topics were included only if we were able to find articles for which we could formulate a useful set of questions. We were pleased that this criterion did permit development of a close-to-fully-comprehensive set of topics.

The book also serves three other purposes, which to many instructors will be as important as the purpose described above. The first is to bring to the student real-world applications of textbook economic theory. Stimulating the student's interest in this way, as our classroom testing has shown, often leads to a better understanding of the economic principles involved. Thus the book can serve to bridge the gap between an instructor wishing to teach history and his students wanting relevance to the real world.

The second is to reinforce the economic theory the students have struggled with in their textbooks. This reinforcement is direct, through application of the theory to specific problems, and indirect, through viewing the economic theory in action in a real-world context. Many instructors feel strongly that only by asking students to work through problems and to explain their thinking clearly will the economic theory be learned. The questions in this book are not easily answered by simply referring to a textbook; students must think and thereby discover what they do and do not know, greatly enhancing the learning process.

The third purpose is to provide a set of supplementary readings useful in their own right. Books of readings supplementing courses in economic principles abound, but the readings usually do not come from newspapers, the major source from which the student is likely to be drawing his economic information in the future. Readings books are designed to provide an additional perspective on textbook theory and to further explain that theory and how it is applied. The articles in this book apply economic theory to contemporary economic issues, illustrating and elucidating the tensions between commonly held social values and goals and the institutions which implement and frustrate them. Reading these articles should impress on students, as it did on us, the strong degree to which reality tempers our theory and colors policy proposals.

Many of the articles extend, as well as support, textbook material commonly used in first-year courses. Because of this it is important that students read carefully the introduction provided for each section. These introductions provide an overview of the economic forces relevant to the topic at hand, and give some perspective to the articles in that section. This overview of the relevant economic theory, along with the foundation provided by a principles text and the encouragement provided by the knowledge that economics is the "science of common sense," should allow students to generate answers to the questions following each article. To aid the students in their task we have included, immediately following the table of contents, a table cross-referencing the sections in this book to some commonly-used principles texts. Even with this help, students might find it of use to refer occasionally to the glossary of selected economic terminology included at the end of this book.

Although this book is intended to be primarily a supplement for principles courses, it could be usefully employed in any course using a principles level course as a prerequisite. Instructors of courses on economic issues or on Canadian economic policy might find this book of particular value. A review of the table of contents will reveal that some sections pursue topics (such as the financing of government spending) which, although not difficult, are not extensively dealt with in principles courses. Some instructors might feel that these sections will be of more value to students in post-principles courses. Anything in the book that is actually difficult is clearly marked as appropriate for advanced students.

The following tables cross-reference most of the sections of this book to specific pages in commonly-used principles textbooks. By referring to these tables, students needing review for any section should be quickly directed to relevant and easily-available background reading. The texts cross-referenced are:

I. Drummond, *Economics: Principles and Policies in an Open Economy* (Georgetown, Ontario: Irwin-Dorsey, 1976)

R. Lipsey, G. Sparks, and P. Steiner, *Economics*, 2nd Canadian edition (New York: Harper and Row, 1976)

P. Samuelson and A. Scott, *Economics*, 4th Canadian edition (Toronto: McGraw-Hill Ryerson, 1975)

D. Stager, *Economic Analysis and Canadian Policy*, 2nd edition (Toronto: Butterworth, 1976)

Section		Drummond	Lipsey	Samuelson	Stager
I-1	and I-2 Supply and Demand	113-119, 226-236	70-87	64-73	25-30, 35-43
I-3	Elasticity	137-145	88-102	351-357	30-34, 38-39
I-4	Corn-Hog Cycle	236-243	108-114 124-126	372-373	380-381
I-5	Costs	146-173	188-194, 202-227	416-420 428-438	285-297
I-6	At the Margin	337-340	248-255, 444-453	420-424	304-306 347-348
I-7	Taxes and Subsidies	337-340	467-471	358-360	54-58
I-8	Government-Fixed Prices	315-321	103-107, 115-119	361-364	49-54 381-384 267-269
I-9	Marketing Boards	251-258, 309-315	103-122	374-389	385-386
I-10	Foreign Competition	173-178	729-740	614-615	209-219
I-11	Imperfect Competition	280-307	290-302	442-456, 467-471	330-332
I-12	Regulated Monopolies	260-279	329-333	459-460	370-373
I-13	Taxing a Monopoly	260-279	467-471	452-456	319-323
I-14	Price Discrimination	289-290	278-283	404-405	327-330
II-1	Consumer Behavior	390-400	519-522, 557-562	184-194	92-99
II-2	Inventory Cycle	427-429, 435-441, 486-489	522-528, 582-583, 591	199-204, 238	104
II-5	Fiscal Policy	432-435, 450-455	546-549, 597-606	205-207, 216-219	166-172
II-6	Monetary Policy	353-381	633-644, 653-692	245-299	181-185
II-9	Stagflation	506-508	810-815	331-335, 337-339	159-166
II-10	Wage-Price Controls	511-516	800-803	335-337	191-194
II-11	Interest Rates and	381-387,	407-410,	301-306,	173-177
II-12	Government Financing	404-407	690-691	547-552	123-124
II-13	International Influences and	74-76, 415-420	697-717, 761-764,	222-227, 292-295,	219-229
II-14	Fixed or Floating Rates?	445-450, 466-481	815-818	584-602, 666-668	

Preface **xi**

Part I
MICROECONOMICS

I-1 Supply and Demand

Microeconomics uses supply and demand curves as its basic tools of analysis. Before students can apply microeconomics to business decisions or to policy problems, they must be skilled at interpreting and using these curves. This first section aims to develop this ability by looking at the supply and demand curve interpretation of some markets which are close to being "perfectly competitive": there is a large number of buyers, a large number of sellers, and a homogeneous product.

Equilibrium in these markets (a price and quantity combination that satisfies both demanders and suppliers) is usually quickly established through price changes, keeping the market at the intersection of its supply and demand curves. Most of the economic analysis surrounding these markets, therefore, is centered on shifts in these curves. Supply and demand curves are derived on the *ceteris paribus* principle: they trace the supply of or demand for something as its price changes, *other things remaining the same*. The demand for television sets, for example, depends on their price, but it also depends on the overall level of income in the economy. As the price changes we move up or down the demand curve; as the level of income changes the entire demand curve shifts to the right or the left.

These shifts in the demand or the supply curves throw the market out of equilibrium, instigating a disequilibrium reaction on the part of the market participants. Both price and quantity adjust, moving the market (in a stable system) to the new equilibrium position. In most textbook examples, the price adjustments are much quicker than the quantity adjustments. The markets in the articles of this section all fall into this category. A later section (Section I-11) examines a market in which quantity adjusts faster than price.

I-1 **A Lumber**

Vancouver *Province*, August 13, 1977

Price of lumber zooms

By PAT JOHNSON

A U.S. home building boom has increased lumber prices in the B.C. market from 15 to 20 per cent in the past 2½ months, and local builders fear prices will continue to climb.

Since U.S. builders are the Canadian lumber industry's biggest customers, accounting for half of its sales, they tend to set prices for the Canadian market as well. U.S. housing starts are expected to peak in the last quarter of 1977 at an annual rate of 1.93 million units.

The exchange rate, giving U.S. buyers a better deal for their dollar, hasn't helped any as far as Canadian buyers are concerned.

"The mills are booked two months solid," said Paul Pannozzo, a lumber purchaser for the Lumberland retail chain. He said what happens then is that buyers are forced to bid higher prices in order to snatch enough material for their needs, before it's sold down south.

Pannozzo predicted that "unless California slows down on home-building," B.C. prices could go up another $20 per thousand board feet by 1978. He said there could be a drop in November when cold weather sets in, but prices are likely to jump again in January.

As an example of the recent price jumps, Bob Swannell, general marketing manager for lumber at MacMillan Bloedel Ltd., noted that Western white spruce 2x4s, considered something of a price indicator in the trade, were selling at $190 per thousand board feet on July 29 but had jumped to $209 as of Friday.

He agreed that it's the U.S. housing boom causing the rapid price hikes here and noted that unlike Canada, the U.S. is building a large number of single-family houses which consume a high percentage of framing lumber, pushing up lumber consumption.

This year, said Swannel, the U.S. is likely to exeed its record year of 1972 for single-family dwelling starts.

What has happened, he said, is that the U.S. boom came at the same time the B.C. forest industry had let production drop somewhwat. With inventory down and demand increasing, prices inevitably started rising.

In addition, he said, the demand came at a time of production drop-offs due to summer holidays and fire closures, especially in Washington and Oregon.

Although U.S. housing starts are expected to continue high through 1977 and into the first quarter of 1978, Swannell said prices might level off once mills are into fall production and stocks build up.

But he said the pricing story could change completely if B.C. runs into labor problems, since there are no new contracts yet in the forest industry.

In Toronto, W.M. McCance, director of research with the Housing and Urban Development Association of Canada, predicted that prices of dimensional lumber, now at a record level of $190 a thousand board feet, will go to peaks of $230 in 1978.

McCance said the association acknowledged that the warning has the potential of being a self-fulfilling prophecy because builders might begin heavy stockpiling which could drive prices up.

But the association's other alternative, McCance said, was to say nothing and risk exposing its members to high prices without warning.

William Procter, forest products analyst for Gardiner Watson Ltd. of Toronto, said "record lumber prices have tended to coincide or follow shortly after the peak in housing starts."

The price rise forecast for dimensional lumber would not apply to such other building commodities as plywood, particleboard and waferboard, which are manufactured almost exclusively for the domestic market, Procter added.

---

1. Does the U.S. home building boom cause a movement along the Canadian demand curve for lumber or a shift in this curve?

2. Has the exchange rate risen or fallen? What does this do to the Canadian supply and demand for lumber diagram? (The exchange rate is the price of Canadian dollars in terms of U.S. dollars.)

3. How would you analyse the reference to Washington and Oregon by using a supply and demand diagram?

4. What have B.C. labor problems got to do with the price of lumber?

5. Using a supply-and-demand diagram, explain how a warning of future lumber price rises could be a self-fulfilling prophecy.

6. Economists often talk about excess demand for something bidding up its price. What statement in the article supports this theory?

I-1 **B Coffee**

Toronto *Globe and Mail*, March 6, 1976

Coffee nations still have upper hand in supply-demand struggle

By GEORGE MCNAIR

A low-key supply-and-demand struggle is continuing between coffee producing countries and processors in the consuming countries, but the producing countries still have the upper hand and retail prices are edging higher week by week.

The processors are not buying heavily for roasting, getting only what they need without building inventory. The producing countries are satisfied to just meet demand at what are already record or near record prices.

All this has been brought about by disastrous frost in a large and important producing area in Brazil and in other South and Central American countries. Then the troubles in Angola and the earthquake in Guatemala have further reduced available supplies.

Before these disasters and difficulties, there was a huge pileup of stocks, considered to be enough to last two or three years. This supply is being meted out as if the scarcity were already here. It is a winning ploy for the producing countries, which hold the high cards.

But it can be a losing game for the distributors and processors in the consuming countries.

The prospect is that if the consumer end of the pipeline is unable to halt the price advances, ground coffee will reach $2 a pound. That point, or even one well below it, will mean sharply reduced coffee consumption which, in turn, will mean serious disruption of the entire coffee trade.

There has already been enough resistance to bring coffee prices back down from a short runup after the Guatemalan earthquake. At that, futures prices set highs up to $1.04 a pound. The present high is about 98 cents, but it had to come down to about 95 cents in reaction to the sharp runup.

Chain stores have played an important part in helping to keep prices down, taking small margins and offering many specials. Consumers have found that if they grind the coffee finer, they can get more cups from a package. Others are drinking more tea, and still others are finding that hot soup is a better value for their coffee breaks.

If the producing countries reduce their business sharply, just to take full advantage of what is so far an artificial scarcity, they may find that when their full supply is restored their market will not be able to handle it and they may never win it all back.

Food producing countries are not in the same category as oil producing countries because oil taken out of ground is gone forever. Food producers must consider what damage they are doing to their markets if prices are excessive.

This year, potato producers have come up against this problem. While there is a recognized world scarcity, there is a limit to what the consumer will pay abroad or at home. Excessive prices two years ago were followed by low prices in the ensuing season because of reduced demand and good crops. This season, demand slowed at a much lower figure than two years earlier.

---

1. Is the upward movement of coffee prices due to shifts in the supply curve, shifts in the demand curve, or both? Defend your answer with explicit reference to information given in the article.

2. Interpret the statement "if the consumer end of the pipeline is unable to halt the price advances" in terms of the characteristics of the supply and/or demand curves.

3. The seventh paragraph describes the dynamic reaction of the price of coffee to the Guatemalan earthquake. Describe this reaction more explicitly, with the help of a supply/demand diagram. ("Dynamic" refers to movement over time; i.e., do prices bounce around a lot and if so why?)

4. What would you expect to happen to the price of tea? Explain why.

5. The example of the potato producers is used to warn coffee producers of long-run drawbacks of their present policy. Explain this danger in your own words and describe how it can be captured in a supply/demand diagram.

I-1 C Meat

Financial Times of Canada, October 11, 1976

Increased pork production will prompt substitution for beef

By Douglas Mutch

Increased meat production last summer resulted in attractive meat prices for the consumer. Beef was available at very low prices.

The downward potential of beef prices is limited. However, hog producers are in the expansion phase of the hog cycle and sharply increased supplies will be available during the coming year.

Last fall, hog prices peaked at more than 90 cents a pound in Toronto and 60 cents in Chicago. These high prices stimulated increased production and marketing levels now reflect this increase.

The U.S. agriculture department on Sept. 22 reported the number of hogs and pigs on farms in 14 major producing states as of Sept. 1.

Third and fourth quarter 1976 production should be 17% above last year. First quarter 1977 production could attain a 20% increase. If producer farrowing intentions are unchanged, second and third quarter 1977 slaughter will be 17% and 10% higher respectively.

These increased pork supplies will keep prices below year-earlier marks at least into mid-1977. Pork will be more competitive to beef and all red meat prices will experience downward pressure.

Higher cattle weights and larger slaughter totals kept beef prices depressed throughout the summer. Low pork supplies resulted in beef being very competitively priced in relation to pork.

Changing

This relationship is changing. Beef supplies are still ample but pork production is rapidly increasing. Pork prices must drop and thus encourage substitution of pork for beef.

Bumper grain crops this year in both Canada and the U.S. will further encourage meat production. When the price of feed is low, meat production costs are cut and heavier animals are marketed.

In recent years, high grain and low meat prices caused many farmers to market grain directly. It was more profitable to sell grain rather than feed it and sell the meat.

Surplus grain supplies in North America this year will reverse that trend. Rather than sell low-priced grain, producers will use it for feed and hope to profit from meat production.

Meat prices over the coming year will continue to fall. Producers are too committed to cut production before next summer.

The beef industry is in the process of culling the herds and should recover before the pork industry. However, beef production cutbacks occurred at a poor time for producers.

First, dairy herd culling has increased beef supplies and forced prices lower. Now, increased hog production will provide heavier competition to beef sales.

Second, lower grain prices are encouraging increased production of all meats. The year ahead looks promising for consumers but very dismal for producers.

Douglas Mutch is a Montreal economist specializing in commodity markets.

---

1. Use a supply-and-demand diagram to explain why "Increased meat production last summer resulted in attractive meat prices for the consumer."

2. Use a supply-and-demand diagram to show, with reference to the hog market, why "These high prices stimulated increased production."

3. Interpret the statement "and marketing levels now reflect this increase."

4. Why is it that if "Pork will be more competitive to beef," "all red meat prices will experience downward pressure"? How would this be explained on a supply-and-demand diagram?

5. Why is it the case that "Low pork supplies resulted in beef being competitively priced in relation to pork"?

6. Interpret the first paragraph after the sub-title "Changing" by means of supply-and-demand diagrams. (Use two diagrams, one for beef and one for pork.)

7. What impact do bumper grain crops have on the supply-and-demand diagram for beef?

8. What influence does the price of meat have on the supply-and-demand diagram for the wheat market?

9. Explain the meaning of the article's title.

10. If you have studied the terms "complements" and "substitutes," explain how this terminology is relevant to this article.

I-2 More Supply and Demand

This section is for more advanced students. It examines three markets that are regularly discussed in the business sections of newspapers but are rarely used as examples in principles texts, even though they are close to being "perfectly competitive." Each of these markets is a little unusual and thus requires more care in the application of supply-and-demand curve analysis.

The first of these markets is the commodity futures market, in which contracts for delivery on specific dates in the future are bought and sold. This market allows gamblers to bet on the future prices of commodities, and permits processors and producers to hedge, or insure, against sudden price changes. Commodity market regulations usually limit the amount by which the price of futures can change in a single day, helping to prevent huge price swings and chaotic trading conditions.

The second of these markets is the foreign exchange market. (Your instructor may want you to postpone examination of this market until your class studies the international sector of the economy; section II-13 of this book focuses on the international sector.) The foreign exchange market is the market in which the value of the Canadian dollar (C$) is determined. Under a flexible exchange rate system, the value of the C$ is determined by the forces of supply and demand. This can be analysed in one of two equivalent ways. We can look at the supply of and demand for C$ by foreigners, or we can look at the supply of and demand for foreign exchange by Canadians. In the former approach Canadian exports imply a demand by foreigners for C$ to buy the exports; in the latter approach they imply a supply of foreign exchange to Canadians, as Canadians are paid for their exports. Imports, and net foreign borrowings (net capital inflows), the other two major sources of supplies and demands in this context, can be analysed in a similar fashion.

The third of these markets is the "money," or bond, market. (Your instructor may want you to postpone examination of this market until your class studies the interest rate; section II-11 of this book looks at the interest rate in some detail.) The "price" of "money" or bonds is the interest rate. There are two basic ways in which supply and demand analysis can be used to determine the interest rate. The first approach, called the liquidity preference approach, is to look at the supply and demand for "money" where money is defined as the sum of currency and demand deposits (chequing accounts). This is not the only definition of money used in this context; Section II-6 examines this problem. The supply of money is determined by the monetary authorities (the Bank of Canada), while our demand for money is determined by our income level as well as the "price" of money, the interest rate. The second, equivalent approach, used in our analysis of the third article of this section, is to look at the supply of and demand for "bonds," where bonds is a catchall for a variety of financial assets; this is called the "loanable funds" approach. The equivalence between these two approaches stems from the fact that both money and bonds are financial assets and there is only one interest rate. If the prevailing rate clears one market (say, the money market) but not the other (the bond market), adjustments will proceed until both markets are cleared.

Although both markets in equilibrium produce the same interest rate, the dynamics of the two markets are different. Because the bond market provides a more realistic explanation of how the interest rate changes in disequilibrium positions, the bond market is usually used to analyse interest rate movements. Do not let the fact that bonds have prices as well as interest rates associated with them confuse you. Because the interest return on a bond is in the form of a fixed-dollar coupon (which is "clipped" from the bond whenever the coupon comes due), there is an inverse relationship between the price of a bond and its interest rate yield.

I-2 A The Futures Market

Toronto *Globe and Mail*, April 10, 1976

New reports send coffee futures up limit

© Dow Jones Service

Coffee futures rose the daily limit and there were unsatisfied buying orders at the end of the day. News that Brazil had purchased 500,000 bags of green coffee from Angola and reports that it was seeking to buy a further large quantity from African producers was the principal price-supporting factor.

Roasters were said to be hesitant to buy green coffee in the cash market because of the continued advance in prices. Dealers were equally cautious of selling to them because of inability to buy back hedge contracts they previously sold in the futures market to protect the price of their physical coffee.

Other price-firming influences were an earthquake in several South American coffee-growing countries, a scarcity of Robusta offerings and a higher London futures market.

Cocoa futures continued strong. Buying stemmed from higher than expected U.S. cocoa bean processing in the first quarter of 1976, compared with a year ago.

The Chocolate Manufacturers Association of America, which covers 90 per cent of the industry, said 67,088 short tons of cocoa beans were processed during the first three months of this year, 39.5 per cent more than a year earlier.

Traders noted, however, that first quarter 1975 cocoa bean processing was the smallest ever recorded for any one quarter.

Grain and soybean futures prices fluctuated in a narrow range. Trading was quiet in anticipation of the Agriculture Department's wheat production estimate for the drought-stricken southwest.

Department officials reiterated their views that the Soviet Union would buy more U.S. grain before the Oct. 1 beginning of a five-year, six million to eight million metric ton trading pact.

Pork belly, hog, and cattle prices closed mostly higher, as did broiler prices. Limit gains in some nearby months reflected cash-market price advances for hogs and cattle and particularly for beef, which have risen about 8 cents a pound in the past two days.

Many traders were taking a breathing spell particularly in cattle, where prices in the past couple of weeks scored one of their sharpest upturns on record.

Maine potato futures were mixed. Some demand for contracts developed after selling that began on Wednesday continued to depress prices. Trading was active.

Lower gold bullion prices in London prompted some selling in precious metal futures. Profit taking after recent sharp price advances also developed.

Copper futures were steady after early declines. Weakness in the value of the British pound prompted buying of copper on the London Metal Exchange as a hedge against the decline in the currency.

The International Tin Council in London said total tin metal in its price-stabilization buffer stocks at Dec. 31 were 20,071 metric tons, up from 11,942 tons at Sept. 30.

---

1. What is a coffee future?

2. What does it mean to say that "coffee futures rose the daily limit"? Does this imply equilibrium in the coffee futures market, excess supply, or excess demand? Explain.

3. The article cites a number of factors that would serve to shift the demand curve or the supply curve for coffee futures. Name these factors and explain why and in what direction they shift these curves.

4. What is the "cash market"?

5. Why would roasters "be hesitant to buy green coffee in the cash market because of the continued advance in prices"?

6. What are hedge contracts?

7. Why would dealers be cautious of selling in the cash market because of inability to buy back hedge contracts?

8. Why should higher than expected U.S. cocoa bean processing lead to buying of cocoa futures?

9. Explain the meaning of "Limit gains in some nearby months reflected cash-market price advances for hogs and cattle."

10. What is profit-taking? What does it do to price? Why?

11. What would expectations of a decline in the British pound do to the British demand curve for copper futures? Why?

12. What are price-stabilization buffer stocks? Does the recent change in the tin buffer stocks imply equilibrium, excess supply, or excess demand in the market for tin? In the absence of the buffer stocks, what would you predict would have happened to the price of tin during the fall of 1975? Explain why.

I-2 B The Exchange Rate

Financial Post, August 23, 1975

Importers worry as the C$ falls toward US96c

By John Rolfe

OTTAWA — Canada's importers appear to be the only people worried about the continuing fall-off in international value of the Canadian dollar.

Over the last several months, the C$ has been working its way down to a present low of about US96c vs more than US$1.04 only last year.

Official Ottawa — as somnolent as ever in August — has little to say about the subject.

Even the Canadian Exporters Association — whose members generally can be expected to profit most from a lower-valued Canadian dollar — seems caught up in the summer drift.

A spokesman says, that in final analysis it is foreign demand for Canada's goods that counts. Until this demand increases, gradual changes in value of the dollar make little difference to export opportunities.

International experts in the department of Finance say the most important factor influencing Canada's soft export performance has been very deep recessions in our major trading partners — the U.S., Europe, and Japan.

The C$ decline is also attributed to the cumulative deterioration in Canada's balance of merchandise trade and in its overall balance of international payments.

Pickup not detected

Imports have stood up well because the Canadian recession has not been as deep as, or concurrent with, recessions felt by our major trading partners.

Officials in the Department of Industry, Trade & Commerce say a continuing fall-off in value of the dollar is a positive element for exports. However, because of time lags in gathering statistics, an export pickup has not been detected yet.

Keith Dixon, general manager of the Canadian Importers Association, Toronto, thinks the C$ will decline even further to perhaps US95c — which worries importers. They would prefer to have the C$ at par with the US$.

"But it is true that our imports overall have been holding up better than our exports," Dixon said.

The lower-valued C$ makes imported goods and services more expensive for Canadians to buy.

Dixon hopes the lower-priced C$ will spur exports. If exports do not begin to increase, pressure could be applied to Ottawa to restrict the volume of imports in order to check deterioration of Canada's balance-of-payments position.

As for the C$ outlook, investment dealers **Wood Gundy Ltd.** concludes that "there is very little in the fundamental, to suggest a permanent weakening of the C$." Wood Gundy suggests that by yearend the C$ could be trading in the US97c-98c range, based on an improving Canadian current account deficit as the U.S. economy recovers.

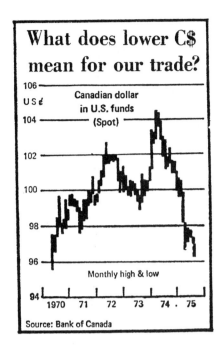

1. Why would Canada's importers be worried about a continuing fall-off in the international value of the Canadian dollar?

2. The fact that the Canadian dollar is changing its value implies that at the time this article was written Canada was on a flexible exchange rate system; i.e., the value of the Canadian dollar was determined by the forces of supply and demand rather than being fixed at a specific level by the government.* As noted in the introduction to this section, these supply and demand forces can be analysed either by looking at the supply of and demand for the C$ by foreigners, or by looking at the supply of and demand for foreign exchange by Canadians. These two approaches are diagrammed in Figure I-2.1 and Figure I-2.2, respectively. Should the exchange rate (on the vertical axis) in these diagrams be measured as Canadian dollars per U.S. dollar or U.S. dollars per Canadian dollar? Explain.

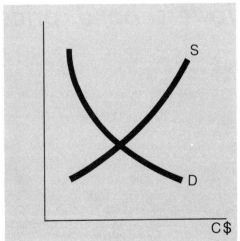

Figure I-2.1

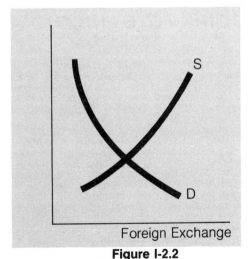

Figure I-2.2

* Actually, Canada was on a "managed" float system in which the government influenced these market forces.

Microeconomics 9

3. Explain what each of the following will do to the supply or demand curves in Figures I-2.1 and I-2.2:

    a) an exogenous (autonomous) increase in imports by Canada

    b) an exogenous increase in exports by Canada

    c) an ad valorem tax on imports (an import tariff) by Canada

    d) an increase in Canadian borrowing abroad.

4. Judging from the contents of the article, explain why the value of the C$ is falling. How can this fall be explained in terms of the supply/demand diagrams?

5. Why would exporters generally "be expected to profit most from a lower-valued Canadian dollar"?

6. What is the Canadian Exporters Association opinion about the price elasticity of the demand for Canadian exports? (Skip this question if you have not yet studied elasticity.)

7. Why should recessions in the United States, Europe, and Japan influence Canada's soft export performance?

8. If the value of the C$ has fallen, why is it the case that "our imports overall have been holding up better than our exports"?

9. Why would the general manager of the Canadian *Importers* Association hope the lower-priced C$ will spur exports?

10. The current account is the difference between exports and imports of both merchandise and invisibles (interest payments and services such as transportation, for example). Why would the Canadian current account deficit improve as the U.S. economy recovers?

11. In the chart presented in the article, the vertical axis measures the C$ in U.S. funds (spot). What does the "spot" mean?

I-2 C The Money Market

Toronto *Globe and Mail* September 4, 1975

Bank rate increase lowers bond prices

By IRVIN LUTSKY

Canadian bond market prices have fallen sharply in response to an increase in the bank rate and prime rate increase announcements by several chartered banks.

Trading was light and the declines, which ranged to almost $2, represented markdowns by dealers. Prices fell sharply from the opening and after stabilizing briefly fell again as chartered banks said prime rates will be increased.

The Canada 4.5s of 1983 were down more than $1 early in the day, but were quoted at $75.50 bid late yesterday, compared with a bid of $77.375 on Tuesday.

Commercial paper rates rose by 25 to 50 cents, with several major borrowers announcing increases on short term paper.

The Bank of Canada increased the bank rate to 9 per cent from 8.25 per cent late Tuesday in a move that surprised dealers. The increase is seen as a signal that the central bank wants interest rates generally higher.

Significant paper issuers such as Ontario Hydro, IAC and Ford Motor Credit of Canada and Traders Group all raised the rates that they will pay on short term notes.

Ontario Hydro raised its rate on 30-day paper to 8.2 per cent from 7.8, and the new 60-day rate is 8.5 per cent, up from 8.2. It pays 8.7 per cent for 90 days, up from 8.4.

IAC increased its 30 to 59-day rate to 8.5 per cent from 8.2 and its rate for less than 30 days is negotiable.

The new Ford Motor Credit rates are 8.375 per cent for one month, 8.5 per cent for two months and 8.625 per cent for three months, all up at least 0.25 of a percentage point.

Traders' new rates for short term paper are 8.5 per cent for 30 to 59 days, up 0.25 of a point; 8.5 per cent for 60 to 89 days, up 0.125 of a point; 8.625 per cent for 90 to 119 days, up 0.125 of a point; 8.625 per cent for 120 to 149 days, up 0.25 of a point; and 8.5 per cent for 150 to 365 days, up 0.125 of a point.

Traders also increased one-to-five-year rates by 0.25 of a point. The new rates are 9.25 per cent fo one to less than two years, 9.375 per cent for two to less than three years, 9.5 per cent for three to less than four years and 9.75 per cent for four to less than five years. The five-to-10-year rate is 10 per cent.

1. The bond market, like other markets, can be represented by supply and demand curves. In contrast to most other markets, however, economists sometimes use the interest rate to represent the "price" of a bond, rather than the actual price of the bond, when drawing these curves. Draw two sets of supply and demand curves for new bond issues, using the interest rate on the vertical axis for one set, and the bond price on the vertical axis for the other. The first sentence of the article states that bond prices fell when the interest rate rose. Are your diagrams consistent with this inverse relationship? Explain.

*2. The article talks about both new bond issues and existing (old) bond issues. There is a market for both. How would the supply and demand curves (use the interest rate on the vertical axis) for the market for existing bonds differ from the corresponding curves for the market for new bonds? In particular, what is the slope of the supply curve for old bonds? And what is the shape of the demand curve for new bonds? (Hint: The demand curve for new bonds is strongly influenced by the existence of a market for old bonds.) What longer-run relationship is there between these two markets? (Hint: Note that once the new bonds are sold they become "existing" bonds.)

3. Would the interest rate in both these markets (the market for new bonds and the market for existing bonds) be the same? If not, why not? If yes, what would cause them to be the same?

4. Explain in words why bond prices fall when interest rates rise.

5. How does the Bank of Canada's action of raising the bank rate affect the supply and demand curves? Explain. Note that the bank rate is used as a bellwether (a signaling device) by the Bank of Canada: A rise in the bank rate signals that the Bank of Canada does not intend to supply as much new money in the future as it did in the past. This means that the chartered banks won't have as much new money to lend (or to invest in the bond market), implying that some of their former customers will have to look elsewhere (i.e., in the bond market) for the funds they want to borrow.

6. The article states that the declines ranged to almost $2. Why wouldn't all the declines be the same? Which bonds would have greater declines and which less?

7. What are Canada 4.5s of 1983? Why are they selling for a price in the range of $78?

8. What is commercial paper? What does it mean to say that commercial paper rates rose by 25 cents?

9. What is a significant paper issuer? What is 30-day paper?

10. Why are the longer-term rates higher than the shorter-term rates? Could the reverse ever be the case?

---

* These questions are more difficult. They can be omitted without loss of continuity.

I-3: Elasticity

The price elasticity of demand is the percentage change in demand caused by a one-percent change in price. It is a mechanical property of demand curves that tells us something about how sensitive demand is to price changes. Given magnitudes of a price change and the corresponding quantity change, elasticity can easily be calculated with a little arithmetic: Divide the percentage change in quantity by the percentage change in price. (The resulting negative sign is often ignored.) An inelastic demand curve is one with low elasticity: a price change elicits very little change in demand. An elastic demand curve is one with high elasticity: a price change alters demand considerably. Note, however, that elasticity really refers to a *point* on the demand curve, not the entire demand curve itself: different points on the same demand curve can and usually do have different elasticities associated with them.

The importance of all this relates to the impact on total revenue of a price change. In the inelastic case, a small percentage price rise causes an even smaller percentage decrease in quantity demanded, so that total revenue, the product of price and quantity, rises. In the elastic case, a small percentage price rise causes a large percentage decrease in quantity demanded; total revenue falls. When elasticity is exactly unity (an increase in price causes the same *percentage* decrease in quantity demanded) total revenue remains unchanged. If demand is inelastic, and thus total revenue rises when prices rise, there is a great temptation to businessmen to raise prices to increase profits. Only competition or government regulation can protect the consumer from being exploited in such markets.

Elasticity of demand (or of supply) can be important for two other reasons. In analysing the dynamics of any market (how it moves from one equilibrium to another), elasticities are important in determining whether or not cycles are generated. For example, a little excess supply might cause a disruptively large fall in price if demand is inelastic. In applying government policy, the success of that policy in terms of its stated goals may very well depend on the relevant elasticities. For example, a lower exchange rate may not stimulate exports if the demand for our exports is inelastic. Both these roles of elasticity should become apparent in later sections of this book.

There are two major problems in measuring elasticity; both appear in the articles in this section. First, elasticity is a *ceteris paribus* concept—it relates to the quantity change resulting from a price change *with everything else held constant*. In the real world, however, we seldom have the opportunity of observing a situation in which price is the only element affecting demand that has changed. Casual elasticity measures, such as the ones that can be calculated from newspaper articles, should therefore be treated as ballpark figures. The second major problem is that elasticity is supposed to be calculated using very small changes in price and quantity, because elasticity may change as we move from point to point along the demand curve. If we have numbers only for large changes in price and quantity, the resultant calculated elasticity will not accurately represent the elasticity at either the old price or the new price. For example, the difference between prices of $10 and $15 could be conceptualized as either a 50 percent rise or a 33 percent fall. Elasticity estimates are usually therefore computed (if possible) using averages of the old and new prices and quantities as the bases from which percentages are calculated, and the elasticity is related to this average position. This measure is called the arc elasticity.

I-3 A Public Transit

Toronto Star, February 21, 1976

FARE INCREASE TO COST TTC 17 MILLION RIDERS

By Ross Howard
Star staff writer

The first effect of Metro's budget cut hits the average Metro resident tomorrow at 9 a.m.—TTC fares jump to 50 cents each or five tickets for $2.

The new fares adopted yesterday are expected to drive off 17 million passengers in 1976, TTC officials said. They are also likely to prompt a major debate at Metro Council on Tuesday.

Starting tomorrow morning the adult fare moves up from 40 cents each to 50 cents and from three tickets for a $1 to five for $2. On March 14 the rest of the increase takes effect: Senior citizens and students, five tickets for $1; children, eight tickets for $1; family and Sunday passes, $1.75.

The fare increase was caused by a Metro Council Executive Committee decision on Tuesday to chop $14 million out of the TTC's 1976 operating deficit of $52 million.

"Metro wants us to cut costs and raise revenues," TTC Chairman Gordon Hurlburt said yesterday. "The only way we could do it is by increasing fares. We don't want to reduce our services."

The loss of 17 million transit riders because of the fare hike will cost the TTC an estimated $14.5 million but Hurlburt said the new fares were set to cover this loss.

The TTC carried 357 million riders last year and was aiming for 366 million this year, before the fare hike was announced. TTC officials said most of the lost riders would be rush-hour users, including senior citizens and students who will now use transit less often and will shop closer to home.

NEW RIGHTS

TTC general manager Michael Warren said "there is a psychological resistance to using transit after a fare increase, but people forget about this over time. If the economy improves and there's more money around, then we'll recover our riders quickly."

Reprinted with permission The Toronto Star.

1. From the information given in the article, calculate the price elasticity of the demand for TTC services.

2. Is the general magnitude of this estimate consistent with the prediction given in the article as to the direction of change of total revenue? Explain.

3. What does the article suggest about the relationship of the short-run demand elasticity to the long-run elasticity?

4. Does the last sentence imply that an improving economy will change the elasticity? If so, in what direction will it change it, and why? If not, why not?

I-3 B Ferry Use

Vancouver *Province*, January 12, 1977

Ferry use declines

VICTORIA (CP) — Figures released by B.C. Ferries Tuesday show that passenger traffic on the system's two main runs declined by 800,000 persons in 1976.

Bill Bouchard, assistant traffic manager with the Crown corporation, said in an interview that 6.6 million people used the Swartz Bay to Tsawwassen and Departure Bay to Horseshoe Bay runs in 1975 but only 5.8 million used them in 1976.

Bouchard said that vehicle traffic on the two runs also declined. Volume in 1975 was 2.30 million compared with 1.97 million in 1976.

The figures show that the Swartz Bay to Tsawwassen run was the system's busiest, moving 3.5 million passengers and 1.1 million vehicles in 1976.

Bouchard said a doubling of ferry fares introduced last year, a general drop in tourist traffic, and bad weather during the summer months were the main reasons for the decline.

He said complete 1976 traffic figures for the system were not available.

---

1. From the information in the article, calculate the elasticity of demand for passenger ferry service and the elasticity of demand for vehicle ferry service. (Hint: Since the original price is not given, let it be an unknown [say, x] for the purpose of calculating the price change as a percentage of the average of the before and after prices.)

2. When the British Columbia government doubled the ferry rates it claimed that revenues would increase. Do your elasticity calculations support this prediction? Explain.

3. Information given in the article suggests that the *ceteris paribus* assumption is not valid. In what direction do you feel your elasticity calculations are biased as a result of this? Explain.

14 Dateline Canada

I-4 The Corn—Hog Cycle

Economic analysis with supply and demand curves can do more than just determine equilibrium price and quantity. It can also aid in determining the nature of the sequential process by which a market moves from an old equilibrium to a new one. Such analyses are called "dynamic." The nature of the dynamic adjustment of price and quantity to a market disequilibrium varies widely, depending on the particular market in question. One very important determining factor is the time lag associated with changing supply. In the agricultural sector this is a critical problem, since supply is usually determined well in advance of its delivery, and cannot be changed until the following year's crop is planted. A second important determining factor is the number of suppliers. When there is a small number of suppliers each individual supplier will take into consideration the expected changes of its competitors when formulating its own supply change. When there is a large number of suppliers, as is typically the case in agricultural markets, a single supplier is not able to gauge the market's reaction, resulting in an uncoordinated supply change. This inevitably leads to an overreaction on the part of the market as a whole, creating an undesirable cyclical behaviour of prices and quantities.

This illustrates one of the shortcomings of the price system: In some markets random shocks lead to cycles which can sometimes be quite severe, and which always cause headaches and heartaches for those trying to make their living in these markets. The government has responded to this problem with policies designed to prevent or cushion these cycles; some of these policies are discussed in later sections of this book.

In economic theory, these kinds of cycles are referred to by the expression "corn—hog cycle," a term that originates from the example first used to illustrate this phenomenon. (The term "cobweb" is also used to describe this phenomenon, since tracing dynamic movements on a supply/demand diagram usually creates a cobweb picture.) Although the article below relates to the corn—hog example itself, the theoretical ideas and economic forces are applicable to a wide variety of markets.

I-4 A Pork Rise Chops Prices

Toronto *Globe and Mail*, June 23, 1976

Pork output to rise; chop in prices seen

WASHINGTON (AP) — Farmers plan to increase hog production more than had been expected in the coming months, meaning more pork and possibly lower prices by early next year, an Agriculture Department report says.

The department's Crop Reporting Board said the baby pig crop is expected to be up 18 per cent this summer and fall from the corresponding period last year. As they grow and are sold for slaughter, those pigs will provide most of the retail work supply during the first half of 1977.

Farmers have cut back sharply on hog production the past few years and officials said the planned increase would restore production to about the 1973 level.

Retail work prices have eased from their record peaks last fall but are not expected to decline much further until the larger hog supply begins to reach the market.

The report said farmers had about 52.6 million hogs and pigs in their June 1 stocks, up 9 per cent from a year earlier but still 11 per cent below the number for June 1, 1974.

In 14 states, which produce about 85 per cent of U.S. pork, farmers indicated they planned to have 2,415,000 sows farrow baby pigs during June-August, up 16 per cent from last summer. In March, department specialists had counted on an increase of about 11 per cent.

Looking to the fall quarter, the report said that farmers in the 14 states indicated they would have 2,506,000 sows give birth this fall, up 19 per cent from September-November last year. It was the first department indication of fall production.

Nationally, the report said a total pig crop of 42.1 million head is expected this summer and fall, up 18 per cent from June-November of last year.

A major reason for the expansion has been that market prices of hogs have been high in relation to feed costs for some months. With last fall's record corn ahrvest and prospects of better profits, farmers plan to feed more to hogs.

The report said the pig crop nationally in the first half of the current marketing year, which began last Nov. 1 and ran through May 31, totalled 41.4 million head, up 16 per cent from the same six-month period last season.

Despite that increase, however, the pig crop in the first half of the year was the second smallest since 1937, officials said.

---

1. Judging from the contents of the article, does the current change in hog production result from a movement along the supply curve, a shift in the supply curve, or both? Defend your answer.

2. Why should pork prices be expected to decline as the larger hog supply begins to reach the market?

3. What do you anticipate will be the size of next year's hog production relative to this year's production? Explain.

4. Does your answer to (3) above suggest a cyclical movement in the price and output of pork? Explain by using a supply/demand diagram.

5. Would a higher demand elasticity for pork (with respect to price) intensify or alleviate the cycle described in (4) above? Explain. (Hint: Draw a graph.)

6. Would a higher supply elasticity intensify or alleviate the cycle? Explain.

7. Would the cycle be intensified or alleviated by recognizing explicitly the role played in the hog market by the corn market? Explain.

8. How would the cycle (described in (4) above) differ if the hog-producing business were controlled by one giant corporation rather than by thousands of small farmers? Explain.

16 DATELINE CANADA

I-5 Costs

The cost and supply dimension of economic problems in newspaper articles can be analysed by applying a few basic principles. The points below summarize these principles.

1. The marginal cost (MC) curve passes through the bottom points of the average variable cost (AVC) curve and the average total cost (ATC) curve. This is true because of the arithmetical relationship between marginal and average costs.
2. These costs include opportunity costs, in particular the return that the firm could have made on its assets had they been employed elsewhere. This implies, for example, that the ATC curve includes a "normal return" or "normal profit" which must be earned to keep the firm in this business.
3. In the short run a firm will continue to operate as long as average variable costs are covered. In the long run it will operate only if average total costs are covered.
4. The long-run ATC curve is the envelope of the ATC curves associated with plants of all possible sizes.
5. Firms in competitive markets maximize profits by producing the output at which marginal cost equals the market-determined price. Thus the portion of the firm's MC curve above its AVC curve gives that firm's short-run supply curve. Summing these curves horizontally over all firms gives the industry short-run supply curve.
6. In the long run, supply is determined by allowing the size and number of firms to change, and factor prices to adjust.
7. Subsidies and taxes levied on producers can be analysed by noting their impact on the firms' cost curves.

Ceteris Paribus Assumption — only when we assume that all other factors (1-6) are constant with only price determining quantity

I-5 A **Growing Cherries** Vancouver Sun, May 31, 1977

Cherry growers lost money

PENTICTON (CP)- Charles Bernhardt, president of the Okanagan Fruit Growers Association, said Monday fruit growers will lose money on the 1976 soft fruit crop despite federal and provincial price assistance.

Bernhardt said in an interview that income from fruit sales combined with government assistance will not meet the real costs of producing the fruit.

But he said he wasn't seeking enough government support to end all losses because "if we were assured of all our costs of production we would lose the necessary incentive and concern for production of quality fruit."

Bernhardt said the only way growers can make a profit is to have high returns from the marketplace because government support programs cover only part of production costs.

Growers, he said, will lose six cents a pound on the 1976 cherry crop. While they cost 37.5 cents a pound to produce, the return from the marketplace and government subsidies will bring in only 31.5 cents.

It is estimated that federal and provincial government subsidies for the 1976 crop will be 13 cents a pound for cherries, 7.7 cents for apricots, 6.5 cents for pears and six cents on plums.

While calling the support levels fair, Bernhardt said he was disappointed that limits will be put on the amount of a grower's crop that will qualify for a subsidy.

He said a grower can get a subsidy on only 25 tons of his cherry crop, but one of his members harvests up to 200 tons.

Bernhardt said what Canadian growers need most is tariff protection from United States and overseas fruit coming into the country.

He said with U.S. harvest dates ahead of Canada, the fruit hits the Canadian market at a time of high prices, and by the time local fruit comes to market the prices have often declined below the cost of production.

---

1. If cherry producers expect to lose 6¢ per pound, why do they bother producing at all?

2. If an unlimited subsidy of 13¢ per pound is given to all cherry producers, what happens to the MC, AVC, and ATC curves?

3. What happens to the MC, AVC, and ATC curves if only the first 25 tons are eligible for the subsidy?

4. Suppose all cherry producers currently each produce at least 25 tons.

   a) What happens to the supply curve if the subsidy is unlimited?

   b) What happens to the supply curve if the subsidy is limited to the first 25 tons?

5. Besides the obvious fact that the limited subsidy will cost less money, does your answer to the preceding question suggest any additional reasons why the government might prefer a limited subsidy?

6. What impact would tariff protection have on the supply-and-demand diagram for domestic fruit?

I-5 **B Copper Smelting**

Smelter not viable

Five mining companies reject plan for B.C.

Separate major studies by five mining companies in the last few years have all concluded copper smelting and refining is not economically feasible in B.C. — and will not be for at least five years.

Doug Little, executive vice-president of Placer Development Limited, said Placer, Noranda Mines Ltd., Cominco Ltd., Bethlehem Copper Corporation Ltd. and Granby Mining Corporation have all looked into the possibility of building a smelter — and have rejected the idea.

In information compiled by the research department of Placer for The Vancouver Sun, Andy McQuire, a marketing engineer, said the cost of smelting and refining copper in B.C. would be 37 cents a pound. This compares with 16 to 22 cents B.C. copper mines now pay Japanese copper smelters to do the job.

"From Placer's point of view, there could be little financial justification for building a copper smelter and refinery complex in B.C.

"Rising costs, depressed markets, labor unrest and government policies towards the environment are major factors which contribute to high ownership and operating costs," McQuire's report said.

These factors do not just operate in B.C., but also apply in general to the Western World smelting situation.

Capital and operating costs for new installations have increased substantially in recent years with the result that old operators are faced with cost pressures and potential new investors have less ability or inclination to invest in smelters," he said.

The report listed nine smelter projects around the world that have been closed or deferred in the last few years. They include two in Canada, three in the United States and one each in Japan, Australia, Mexico and the Philippines.

This would seem to work in favor of new smelters — if no one else is building, whoever gets in first will clean up.

But this is not the case. There is already a world oversupply of smelting/refining capacity, the report said, and there will be for at least five more years.

The report said that in 1974 Western World smelter/refineries operated at an average 89-per-cent capacity and in 1975 at 81 per cent. They are projected to operate at 83-per-cent capacity this year, 85 per cent in 1977 and 88 per cent in 1978.

Jack Butterfield, head of marketing for Placer, said overcapacity is the fundamental problem in smelting-refining now.

And he said, "The single most important factor in over-capacity now is that in both consuming and producing countries governments have influenced companies to build smelters."

He said governments perceive an advantage in having the 16 to 22 cents a pound smelting charge spent within their borders for the jobs and revenue it creates.

As a result, he said, some of them are not concerned that their smelting and refining operations are uneconomic. This puts an additional burden on the independent company trying to run a smelter for profit.

Little commented, "The mining industry in B.C. will do what's economic, but today costs are such that if a smelter had to be built some mines in the province would have to shut down."

Because of the higher cost of smelting-refining in B.C. the copper mines would in effect have to subsidize a smelter if one were built here, Little said.

Marketing the sulphur by-product of a smelter-refinery would also be a problem, Butterfield said.

A 100,000-metric-tons-a-year copper plant would have a by-product of 300,000 metric tons of sulphuric acid.

The market for sulphuric acid has been poor for the last 15 to 20 years, Butterfield said, and it is unlikely to improve. The acid has to be stored, and cannot be easily dumped. "You can't even give it away in a bad market," he said.

McQuire said the high cost of labor is another problem in B.C. "Canadian labor rates are in general higher than in most other countries and B.C. labor costs in the mining industry may be among the highest in the world." Labor disruption is also a problem. For example he said the B.C. Railway, which ships mineral concentrates, was shut down three times this year.

"The cost of money in Canada is higher than in most other countries, say nine per cent annually compared with seven per cent in the United States and six per cent in Japan. In a capital intensive investment, this interest cost difference becomes a major factor."

McQuire summed up, "All of these factors affect profitability, and together explain why as much as 37-cents-a-pound copper smelted and refined in B.C. would be required as a tolling fee.

"Almost without exception, rather than pay this fee, the B.C. mines are prepared to pay the transportation cost required to ship concentrate for processing abroad, where fees are generally 16 to 22 cents a pound."

The Placer research department also provided a breakdown of the costs involved in running a smelter/refinery in B.C.

The report assumes the capital cost of a 100,000-metric-tons-a-year smelter-refinery at $200 million, or $2,000 an annual ton. Operating costs would be 11 cents a pound, income tax rate 45 per cent, complex operating life 15 years, depreciation $133 an annual ton or 6.6 cents an annual pound and return on investment would be 15 per cent. The report did not include provision for any sort of subsidies.

On the basis of these assumptions it would cost 26 cents a pound smelted to recover the initial capital involved, pay taxes and account for the profit. When this is added to the 11-cent-a-pound operating cost the total is 37 cents a pound.

1. In textbooks, costs are usually discussed by drawing curves illustrating cost figures for a range of outputs. The numbers presented in this article refer to a specific output, 100,000 metric tons/year. This output was probably determined by the firm estimating the amount of business it would get were the smelter in operation and pricing competitively. The figures relate to a specific plant size, which we, for convenience, refer to as S. How did the firm determine the plant size? (Hint: Pay particular attention to the difference between short- and long-run cost curves.)

2. For plant size S, is the output level, 100,000 metric tons/year, associated with the bottom of the average cost curve or some other point on the average cost curve? Explain your answer.

3. From the figures presented, for which of the following can you determine numbers: fixed cost, average fixed cost, variable cost, average variable cost, marginal cost, total cost, average total cost? Calculate those you claim are possible to calculate; for the others explain why you are unable to calculate them.
NOTE: A metric ton is approximately 2200 pounds.

4. The article states that "government policies towards the environment" have something to do with the firm's cost function. Explain this, noting whether it affects fixed or variable costs.

5. In the next-to-last paragraph, in which cost figures are discussed, why is the return on investment (or profit for the company) included as a *cost*? Relate this to your textbook theory.

6. Suppose transportation costs associated with having the smelting done in Japan at 22¢ a pound are 10¢ a pound. At what price of sulphuric acid would the domestic smelter become viable?

7. There is considerable debate in macroeconomics concerning the influence on investment of the interest rate. Do the contents of this article offer any evidence supporting either side of this controversy?

8. Evaluate the statement "governments perceive an advantage in having the 16 to 22 cents per pound smelting charge spent within their borders."

9. The article indicates that Japan gets the Canadian business, suggesting that in the world excess smelter supply situation Japan is surviving. This could be because of government subsidies. It could also be due to the cost structure of businesses in Japan. In particular, Japanese firms use more debt (rather than equity) financing than Western firms, and hire labor on the understanding that they will be with the firm for their entire careers (no layoffs). Explain how this different cost structure could lead to Japanese survival in excess supply situations. (Hint: Consider the implications of this cost structure for the breakdown of costs between fixed and variable costs.)

I-6 At the Margin

One of the basic principles of microeconomics is that decisions should be made at the margin. This means looking at the costs and benefits of a small change in quantity and calculating the desirability of that change on the basis only of the extra costs and benefits involved, ignoring overall profitability or desirability of the operation as a whole. The adjective "marginal" is used to capture this idea of *extra* costs or *extra* benefits due to a unit quantity change. The change should be undertaken if marginal costs are less than marginal benefits, since more will be added to total benefits than will be added to total costs. When marginal costs exceed marginal benefits, quantity should be decreased. The "right" quantity is therefore the quantity at which marginal costs equal marginal benefits. Marginal analysis, then, consists of finding the quantity for which marginal costs equal marginal benefits.

Why can the overall profitability or desirability be ignored when contemplating these changes? The reason is that by equating marginal costs and marginal benefits, we will automatically be maximizing the difference between overall costs and overall benefits. This is not to say that the difference between overall benefits and overall costs will necessarily be positive; once we have moved to a position in which marginal cost equals marginal benefit, the overall picture must be checked to ensure that overall benefits exceed overall costs.

This procedure of equating marginal benefits and marginal costs is a valuable tool in any economic analysis involving maximization. The typical example is a firm maximizing profits: profits are maximized when the firm produces the output at which marginal cost equals marginal benefit, where in this instance marginal benefit is more appropriately called marginal revenue. Another example is the consumer maximizing utility: utility is maximized by buying the quantity of the good at which its marginal cost (its price) is equal to its marginal utility.

The most interesting cases of application of this marginal principle, however, occur when conflicting goals are being maximized by business and government. Business wishes to maximize its profit and will therefore equate its marginal cost with its marginal benefit. Government wishes to maximize society's overall welfare and will therefore want business to equate society's marginal cost with society's marginal benefit. The problem is that the firm's marginal cost isn't necessarily the same as the marginal cost to society and the firm's marginal benefit isn't necessarily the same as the marginal benefit to society. For example, marginal costs may differ because of external diseconomies: pollution costs may be borne by society and not by the firm. As another example, marginal benefits may differ because the firm has little competition and faces a downward-sloping demand curve. Society's marginal benefit is measured by the price, but the firm's marginal revenue is measured by the price *less* the loss in revenue on existing output due to the fall in price caused by the sale of the extra unit of output. In examples such as these the government feels justified in taking policy action to adjust the operation of the market mechanism. These cases will be made clearer and illustrated in later sections of this book.

If external economies and diseconomies* are ignored, the relevant marginal benefits and marginal costs to society are given by the demand and supply curves, respectively. This is so by the definitions of the demand and supply curves. At a given price the maximum amount demanded is that quantity at which price equals marginal utility: at a smaller quantity marginal utility would exceed price and the consumer would gain by demanding more, and at a larger quantity marginal utility would be less than price and the consumer would gain by demanding less. At a given price the maximum amount a firm will produce is that quantity at which marginal cost equals that price—it could increase its profits by changing quantity if that were not the case. This use of the demand and supply curves is illustrated in the second article below.

* External diseconomies are costs, such as pollution, that are not levied on producers. External economies, such as smells from a bakery, are benefits for which the producers are unable to charge.

I-6 A Air Cargo Operations

Toronto *Globe and Mail*, April 20, 1976

Air cargo operations believed unprofitable

By KEN ROMAIN

Air cargo operations by Canadian airlines do not appear to be meeting their costs and in general do not appear to be profitable, according to a study of air cargo activity by the Canadian Transport Commission.

Because there is no uniform procedure for establishing the costs of all-cargo aircraft operations and mixed passenger-cargo operations, the study suggests that passenger fares may be paying part of cargo costs.

The airlines in general make no attempt to establish the cost of carrying air cargo in the under-floor space of passenger aircraft, the study says.

"Cargo revenues are simply deducted from total operating expenses to arrive at the cost for passenger services, thus perhaps implying that total cargo costs equal cargo rvenue," which may not be the case.

For comparable distances, the cargo rates on domestic transcontinental routes are considerably lower than on trans-Atlantic routes.

"While the international rates appear to be well above the fully allocated costs of air cargo, the domestic rates seem to cover marginally, if at all, the cost of air cargo."

An analysis of air freight rates for general commodity cargo, the principal category, shows that fully allocated costs of all-cargo services are only marginally covered by the rates.

When lower rate categories are applied, such as specific commodity rates, it would appear that all-cargo services are not recovering costs and many short-haul rates appear to be below costs.

Air Canada's container rates, which are lower than general commodity rates, on domestic transcontinental routes also appear to be consistently lower than the cost of the service.

The study says air cargo makes an important contribution to the revenue of Canadian air carriers. In 1973, cargo revenue for Air Canada, CP Air and the five regional carriers totalled $128-million, 14 per cent of total revenue.

Although the use of all-cargo aircraft is steadily increasing, cargo is still predominantly carried in the belly compartments of passenger aircraft. However, airlines are using only about 20 to 30 per cent of this belly capacity, an extremely low figure.

Transcontinental routes are an important source of cargo revenue for both Air Canada and CP Air, providing Air Canada with 33 per cent of its total cargo revenue in 1973 and CP Air with 21 per cent. CP Air's cargo capacity is limited to 25 per cent of the capacity provided by Air Canada.

Canadian carriers are faced with the problem of traffic imbalances, which also increase costs.

Historically, about three times as much international cargo traffic comes into Canada from the Orient as moves out from Canada. For the North Atlantic, the ratio is two to one in favor of inbound traffic.

Ground handling is a major factor in cargo operations and accounts for about 20 per cent of over-all costs.

However, ground costs are greatly reduced when shippers or freight forwarders consolidate shipments or place cargo in containers.

One reason that there has not been much progress in the development of intermodal container shipments by air is the high empty weight of surface containers, which imposes payload limitations on freight aircraft.

1. Would it be fair to evaluate the cargo operations of Canadian airlines in terms of a marginal increase in output? Why or why not?

2. What evidence in the article suggests that the Canadian Transport Commission does not consider that cargo operations should be analysed in this way?

3. In what direction would a profitability study of cargo operations be biased if these operations were not considered in marginal terms? Explain why.

4. What role do opportunity costs play in this question of cargo profitability?

5. In the sixth paragraph, the word "marginally" is used. Is this the same "marginal" that is discussed in the introduction to this section? If yes, explain the meaning of that paragraph. If no, explain its meaning in that paragraph.

6. If you were president of Air Canada or CP Air, what would you do upon reading this article? Explain.

I-6 B Electric Metering

Toronto *Globe and Mail*, April 10, 1976

Time of day electric metering may be coming

By EDWARD CLIFFORD

The advent of "time-of-day" electric metering, particularly for industrial and commercial power customers, may not be far away, according to the experts.

Such meters not only measure the amount of electricity used by a customer, but when it is used. The customer is subsequently charged a higher amount for power consumed during high-demand hours.

If this system is practiced widely, and providing the price difference between peak power and off-peak power is significant, power consumption can remain at a relatively even level instead of alternating between highs and lows, on both a daily and seasonal basis.

This means less generating capacity needs to be built to meet peak demand, and less fossil fuel is required to fire up thermal stations.

Ontario Hydro expects to make submissions later this year to the provincial Government on cost and pricing, and these could include recommendations to introduce time-of-day costing, according to Gordon Davidson, director of costing and pricing studies for the utility.

The immediate applications appear more likely in industrial and commercial sectors, in which customers are relatively few and demand is high.

Among domestic customers, however, the cost of installing new meters outweighs, for the present, the savings that might result from switching domestic power demand to off-peak hours. It is estimated that a meter with a time and double tariff relay would cost about eight times more than the standard domestic meter.

The model for such a metering plan that is being examined by Canadian utilities is the town of Rutland, Vt., where the two-price system has been introduced.

Off-peak power is sold at one-tenth the cost of peak power to encourage consumers to use their stoves, washers, dryers, water heaters and irons at times that do not coincide with high industrial and commercial demand.

(In the Netherlands and Switzerland, however, where about 50 per cent of all meters are the timing type, the price differential between the two power rates ranges from two to one to four to one, according to B. W. Krutina, president of Landis and Gyr Inc. of Montreal, a meter manufacturer of Swiss parentage.)

The Rutland experiment encourages consumers to use off-peak power by means of a red light in a prominent location inside the house, which shows when there is peak demand. The light is controlled by an operator in the power station who can switch whole blocks of homes into a peak or off-peak condition.

This eliminates the need for a timing device on meters, although the double tariff relay remains. Meters with this equipment alone cost only 50 per cent more than the standard meter.

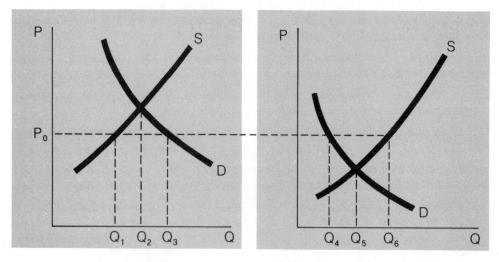

Figure I-6.1a **Figure I-6.1b**

Suppose the current situation can be depicted as in Figure I-6.1a, representing supply (marginal cost) and demand (marginal benefit) in peak periods, and Figure I-6.1b, representing supply and demand in off-peak periods. The supply curve is the same in both periods because the same generating facilities are used. Suppose the current price is a uniform price P_0 for both peak and off-peak electricity.

1. What statement in the article suggests that the marginal cost of electricity is rising, and is probably above average cost during peak hours?

2. From the diagram, what quantity of electricity will be generated in each period?

3. Does marginal cost equal marginal benefit in each period? If not, use the diagram to show whether marginal cost exceeds marginal benefit or vice-versa.

4. Assuming that the demand for off-peak power is not affected by the price for peak power and that the demand for peak power is not affected by the price of off-peak power, explain clearly how society can benefit from charging different prices for peak versus off-peak power. What specific solution would you suggest, in terms of the diagrams?

5. How would you measure the net gain to society? Be explicit, by explaining how this gain can be measured on the diagrams.

6. The two diagrams are not independent, since a change in the price of one kind of power will affect the demand for both types of power. How would this phenomenon best be captured in the diagrams? How would recognition of this affect your answer to question (4) above?

I-7 Taxes and Subsidies

No one denies that the price system—the use of prices to allocate and distribute scarce goods and services—has flaws. There are three major flaws that receive considerable attention from policy makers. The first is that sometimes the dynamics of the operation of the price system creates unnecessary (and undesirable) cycles. The second is that the income distribution created by the system is thought by many to be "unjust." The third is that often not all costs are borne by the suppliers (external diseconomies) or not all benefits are paid for by the consumers (external economies).

A variety of government policies have been directed at these problems. Examples are price floors, price ceilings, marketing boards, fixed prices, taxes, subsidies, and various kinds of income redistribution schemes. Of all the interferences in the market mechanism that have been proposed by governments, the use of taxes and subsidies is thought by economists to be the least unpalatable. This is because the use of taxes or subsidies still allows allocation and distribution to be accomplished by the price system; all the taxes and subsidies do is alter the prices to what the policy makers feel are the "correct" prices.

An example often used to illustrate this is the case of pollution, an external diseconomy. Pollution is a "diseconomy" or cost to society since it detracts from our well-being. This cost, however, is "external" to the firm doing the polluting: the firm does not bear the cost of the pollution. Thus the cost of producing the good in question consists of the normal input costs, borne by the firm, plus the cost of the pollution, borne by society. Because the firm does not bear all of the costs associated with production of the good, the profit-maximizing quantity of the firm will involve a net loss to society in the form of excessive pollution. By taxing the firm's output, the government can force the firm to internalize the pollution cost and thereby cause them to change their profit-maximizing output level, eliminating this net loss. A more explicit example should make this clearer.

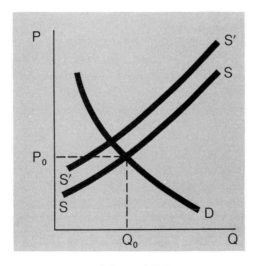

Figure I-7.1

Suppose the supply and demand curves for this good can be shown as in Figure I-7.1. But suppose that firms pollute the air during their production process. To society this is a cost, but this cost is not included in the supply curve, since the supply curve reflects only costs internal to the firm. Thus the supply curve does not incorporate all of the costs to society of producing that good. At the quantity Q_0 the price P_0 reflects the benefit to society of consuming the last unit of output (i.e., from the meaning of the demand curve), but it does not reflect the total extra cost to society of producing that last unit of output, since the pollution cost is ignored. If the government imposes a tax on the firm's output, shifting the supply curve to what it should be (S'S'), from society's point of view, the resulting price and output will be "correct." Here "correct" means that the total extra cost to society of producing the last unit of output is equal to what society is willing to pay for that unit of output (the benefit to society of that last unit). If this equality did not hold, society could gain by changing the output level. If, for example, the cost of producing the last unit of output were greater than the benefit derived therefrom, society could gain by reducing output (costs would be reduced by more than benefits). If the cost of producing the last unit were less than the benefit derived therefrom, society could gain by increasing output (costs would be increased by less than benefits).

This illustrates a general use of taxes and subsidies—to force incorporation of external economies and diseconomies into the supply and demand curves, leading to the "correct" output levels. The pollution example given above is incomplete since it does not recognize the possibility of applying pollution-control devices; the extension of the analysis to include this possibility uses the same principle of applying taxes to make the market move by itself to the correct output level.

Because many different kinds of taxes and subsidies could be applied in a given instance (lump-sum taxes, per unit taxes, ad valorem taxes, and taxes on profits, for example) economic analysis must determine the most appropriate kind of tax or subsidy as well as the appropriate magnitude. The articles in this section allow examination of this problem.

I-7 A Non-Refillables

Toronto Star, March 15, 1976

Non-refillables are a lot of garbage

Sometime this week Ontario Environment Minister George Kerr has to say what he's going to do about non-refillable pop bottles and cans. The deadline the Ontario government set for the industry to voluntarily increase the availability of returnable bottles expired on Saturday. Now Kerr has to decide whether a voluntary approach has worked.

The Ontario Soft Drink Association has tried to encourage the public to return bottles by doubling the deposit consumers get back. And it has increased the availability of refillable bottles. But the industry still has a long way to go. According to the environment department's Waste Management Advisory Board, more than 60 per cent of the soft drinks sold in Ontario are still in non-refillable bottles and cans, which isn't much better than a year ago.

Aside from the fact that non-returnable bottles and cans are wasteful, they are problems because they add to a mushrooming volume of garbage that big cities are finding increasingly difficult to dispose of. They also are a major cause of litter and broken bottles and cans may lead to nasty accidents.

The best answer would seem to be to phase out the use of non-refillable bottles and to impose a special tax on pop cans. Beer drinkers already have to pay more if they want cans.

This could not be done overnight because the bottling and canning industries have big investments in the existing system of packaging. But it should become the target for Ontario, to be achieved in, say, five years.

Pollution Probe estimates consumers would save close to $8 million a year, parks and roadways would be cleaner and we would have done something to level off the growing mountains of garbage that must be disposed of every day of the week.

The way we live is changing. It means we will probably have to change our attitudes about many things, some of them as mundane as pop bottles. After all a pop industry based on non-refillable containers probably uses more energy than one that doesn't, as well as wasting natural resources.

Kerr should keep these things in mind—as well as the fact that consumers will save money—when he draws up his statement on the future of pop bottles and cans in Ontario this week.

Reprinted with permission The Toronto Star.

---

1. Use a supply/demand diagram (assuming the price of pop in returnable bottles constant) to show the influence on the consumption of pop in non-refillable bottles of:

    a) a per unit tax on non-refillable bottles

    b) a sales tax on pop purchases in non-refillable bottles.

2. What influence would an increase in the price of pop in returnable bottles have on the supply/demand diagram in (a) above?

3. Under what circumstances, in question (1) above, is the tax passed fully to the consumer? Under what circumstances is it passed fully to the producer?

4. Does it matter to your answer in (3) above, whether the tax is collected from the consumer or the producer? Explain.

5. The environmentalists complain because producers of non-returnables do not bear the cost of the externalities (garbage and litter) caused by their product. Imposing a tax on the non-returnables is designed to raise their money cost to cover the total (private plus social) cost. Banning them is an alternative solution. The first solution creates an "optimal" amount of pollution (garbage and litter) whereas the second solution creates zero garbage and litter, a less than "optimal" amount.

a) What is the meaning of the "optimal" amount of pollution?

b) Using the supply/demand diagram from (1) above, show how the "optimal" amount of non-refillable bottles can be determined.

c) In terms of your diagram, what is gained by adopting the tax solution rather than the banning solution? Rather than a "do-nothing" solution?

I-7 B Surplus Milk

Toronto *Globe and Mail*, March 9, 1976

Who financed the surplus?

The Canadian Dairy Commission is an agent of the federal Government. It uses public money to subsidize farmers for the industrial milk they produce for, among other things, butter and skim milk powder. It uses public money to back up its contractual obligations to buy, at a fixed price set by the Government, all butter and milk powder produced in Canada surplus to market demand.

And Canada's dairy farmers are saying that the cost of their numbing overproduction of industrial milk in the 1975-76 dairy year (April to March) — an overproduction now measured in surpluses exceeding 300 million pounds of milk powder and 50 million pounds of butter — is borne only by them and not by the public.

It is an argument that does seem solid on the surface. But underneath . . .

The CDC subsidy for industrial milk, set for 1975-76 at $2.66 a hundredweight, is tied to the projected demand for the year. Thus, with a projection set at 100 million hundredweight, CDC's total subsidy pie—which is a fixed figure, not altered by what actual production turns out to be—was $266-million plus $9-million for storage and marketing of normal seasonal surpluses.

When production skipped past demand —by at least 11 million hundredweight for the year, with production increases of 20 to 26 per cent in the second half of the year over the comparable period for 1974-75—the subsidies the farmers received did, in fact, shrink.

The total CDC pie remained unchanged but money had to be diverted away from subsidies to cover the CDC's "constant offer to purchase" all the hundreds of millions of pounds of surplus milk powder and butter that the market could not absorb.

Money would have to cover the interest on loans to the CDC from the public treasury (the Government has set aside $300-million) so that it can meet its purchase obligations. It must cover storage of the surpluses (a cost now of $2.5-million a month with a conservatively estimated total bill for the year of $20-million, well beyond the budgeted $9-million).

It must cover the losses the CDC will incur by having to buy at its own established "floor price" of $1.03 a pound for butter and 64 cents a pound for milk powder with the grim expectation of being able to resell the butter for only a third of the floor price and the milk powder, at last look, for between 14 and 5 cents a pound.

So in October and November, for example, farmers got only $2 a hundredweight subsidy on deliveries and for December it was $1.86. Moreover, the levy assessed against each farmer to further cover costs of storage and marketing of CDC-purchased surpluses was increased from 45 cents a hundredweight at the start of the dairy year to 90 cents by July 1 and now—currently—65 cents.

So *quod erat demonstrandum:* the costs of overproduction are coming out of the farmers' own pockets.

Not entirely, no.

There is the $10-million from the Canadian International Development Agency to buy milk powder at floor prices for foreign aid. There is the 34-cents-a-pound so-called "consumer subsidy"—the subsidy that always seems to be forgotten—on milk powder sold for human consumption which, of course, artificially boosts market demand. And apart from these kinds of items there is the biggest subsidy of all—the one that is putting the money into the farmers' pockets to cover their losses.

The price of industrial milk and its products is totally controlled by Ottawa —which sets the "target support price" that the farmer will receive and the "floor price" that the processor will receive . . . in effect, setting the prices that the consumer will pay.

And the Government, as Agriculture Minister Eugene Whelan was boasting at a recent gathering of dairymen, has driven up the price of industrial milk by 30 per cent since April, 1974—a price increase which the Food Prices Review Board found had little to do with increased costs but was intended primarily to improve the dairy farmers' incomes. Moreover, the federal Government has given the dairymen a formula for future price increases that places less than half its weighting factors on increased costs and will lead to more public subsidies paid out at the grocery store.

Mr. Whelan was canny to warn the dairymen that Canadians will not tolerate the kind of surpluses they produced last year. He might have gone farther and warned them that Canadians are unlikely to tolerate much longer paying for the monopolistic cartel the Government has handed the industry.

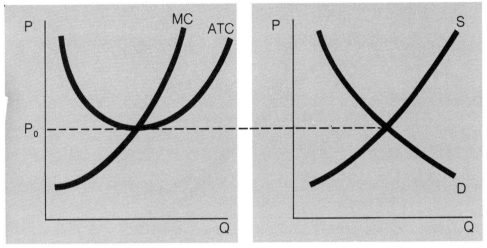

Figure I-7.2 **Figure I-7.3**

1. Suppose the cost situation of a typical milk producer can be represented by Figure I-7.2, where the price P_0 is given by the competitive market, pictured in Figure I-7.3, and the producer produces where marginal cost (MC) equals price. Average total cost (ATC) includes a return or "profit" to the farmer, just sufficient to keep him in the business. Explain the effect of each of the following subsidy schemes on Figures I-7.2 and I-7.3, and explain their impact on price and quantity.

   a) A subsidy to each existing farmer in the form of a doubling of the "profit" or "return" required to keep him in business.

   b) A subsidy to each existing farmer in the form of a fixed per capita grant.

   *c) A subsidy to each existing farmer in the form of a per-unit subsidy. (Hint: The subsidy changes cost curves.)

2. How would your answers to (1) above be changed if the subsidies applied to new entrants as well as existing producers?

3. For each of the cases in (1) above, who receives the benefit of the subsidy, the producers or the consumers? Explain.

4. Suppose a price floor is imposed by the government with a guarantee to purchase anything offered at this price. What effect does this have on Figure I-7.3?

5. The government action described in the article is best analysed as a per-unit subsidy (it certainly appears to the producers, initially, as though it is a per-unit subsidy) combined with a price floor and a government promise to purchase at the floor price. Assume a price floor of P_0 and no new entrants permitted. How would this be described in terms of the diagrams? Does your answer to part (c) of question (3) above change? Can you determine the magnitude of the total subsidy paid on your diagram?

6. Of all the subsidy schemes discussed above, which do you think would be "best" for Canada? Defend your answer.

7. Let us turn now to the "consumer subsidy" of 34 cents per pound, mentioned in the latter half of the article. What does the imposition of such a subsidy do to Fig. I-7.3? Who receives the benefit of this subsidy, consumers or producers? Explain.

---

* For more advanced students.

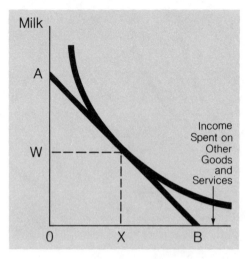

Figure I-7.4

*8. Just as a "cost curve" diagram can be used to analyse producer supply, an "indifference curve" diagram can be used to analyse consumer demand. In Fig. I-7.4 the line AB represents the consumer's budget constraint, so that he/she currently purchases OW of milk and OX of other goods and services.

a) How is the impact of the "consumer subsidy" shown by using the diagram? What happens to the quantity of milk purchased? Show this explicitly.

b) How can the technique used in (a) above be used to derive the demand curve for milk?

c) How can you tell if demand is elastic or inelastic? (Hint: Watch what happens to demand for other goods and services.)

d) Suppose the government, instead of giving a "consumer subsidy" of 34¢ per pound, gives a cash gift to each consumer in an amount sufficient to move him, at the original price of milk, to the combination purchased in part (a) above. Using your diagram from part (a), show which subsidy action would be preferred by the consumer. Which would cost the government more?

* Appropriate for advanced students.

I-8 Government-Fixed Prices

A common government policy is to fix the price of a good or service. This action runs counter to all economic theory which states that only through freely fluctuating prices can the price system allocate and distribute goods and services efficiently. Sometimes the government fixes the price mainly to stabilize it, as is the case in the example of a fixed exchange rate. But more often the price is fixed because the government feels the market-determined price is either too high or too low. The first article in this section looks at the wage rate (the price of labor), a price often thought by governments to be too low. The second article examines the price of rental housing, a price often thought by governments to be too high. The third article discusses the oil market.

A very successful investment counsellor once revealed that his major strategy was "betting against the government": He simply found a market in which the government was attempting to fix the price of something (for example, gold) and then invested on the supposition that the government would be unable to maintain that price. This is one major problem with government price-fixing: usually the economic forces are too strong for the government to contain. The second major problem associated with price-fixing stems from economic theory. With a price fixed either too high or too low (relative to the market-clearing price), allocation and distribution of goods and services will be inefficient—the economy will not be as well off as it could be. If the price of a resource such as labor is fixed too high, not enough labor will be employed, resulting in underutilization of available labor. When the price of a good is fixed too high, too many resources are allocated to the production of the good in question, resulting in unwanted surpluses (some of the resources should have been used to produce other goods in stronger demand). When the price is fixed too low, not enough resources are allocated to its production, resulting in shortages (and rationing if imports are not available).

This misallocation of resources is sometimes pictured by using the concepts of consumers' and producers' surpluses. Consumer's surplus is the difference between what a consumer would be willing to pay for the quantity bought and the amount actually paid for that quantity. The amount he would be willing to pay for the quantity bought can be conceptualized by pretending the consumer were charged the highest price he would pay for the first unit, then the highest price he would pay for the second unit, and so on. Graphically this is given by the area under the demand curve to the left of the quantity purchased. Thus consumer's surplus is given by the triangle WXZ in Fig. I-8.1. Producer's surplus is the difference between revenue and costs of production, shown in Fig. I-8.1 by the triangle XYZ. Although the concept of consumer's surplus is not held in high regard by most economists (they don't like anything that pretends to quantify utility), it is often useful in illustrating misallocation of resources and the effects of government policies.

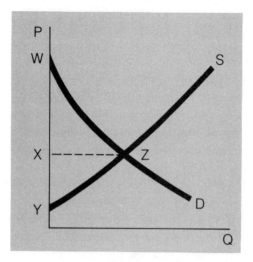

Figure I-8.1

I-8 A **Minimum Wages**   *Financial Times of Canada*, May 2, 1977

Sense and minimum wages

Often, too much is expected of the minimum wage as an instrument of social and economic policies. In fact, not enough is known for sure about its impact on jobs, location costs and investment.

That is all the more reason for provincial and federal governments to reassess the way in which they decide increases. Later this summer the rate could rise at the federal level depending on the result of a review now in progress. And it may well rise in Quebec later this year — again depending on the outcome of a decision by the Quebec cabinet this week to tie its minimum wage increases to a special index. Quebec's plan adds even more urgency to the need for more federal-provincial analysis on variable rates of minimum wage across Canada.

Traditionally, minimum wages are increased by cabinet decision after a review of factors including increases in prices, and hourly or weekly earnings. Regrettably, less attention is paid to the competitive implications of a decision.

For example, six years ago the minimum wage in Quebec and Ontario was $1.50 an hour. Today, Quebec's rate is $3 an hour — 35 cents higher than Ontario and the difference could widen this year under Quebec's plan to index increases to the increase in average hourly wages. (Ontario has already resisted a rise in its minimum rate.) In the U.S., the minimum wages of all the border states is either at or below the federal rate of $2.30 an hour, which under a Carter administration proposal could rise to $2.50, still well behind Quebec's.

High unemployment regions cannot afford to make themselves more uncompetitive in the search for new companies and investment. While low-wage workers are found, typically, in tourism, retail trade and personal services — where companies are unable to relocate — Quebec and the Maritimes should recognize that any increase in minimum rates will also raise wages in industries which pay slightly more than basic rates.

If Quebec expects to attract more small and medium size businesses, which often turn to low-wage markets for labor, the province should try to bring itself back on side, relative to Ontario where the jobless rate is lower.

Unfortunately, Quebec may only add to the differences by opting for an index which will raise the minimum wage regardless of economic conditions and rates in other regions which compete for industry.

However distasteful it may seem, high unemployment provinces cannot ignore lower wage rates in competing regions unless new incentives can be developed to offset the wage advantage.

---

1. Draw a supply and demand diagram of the labor market with "number of people" on the horizontal axis and "wage rate" on the vertical axis. Assume that the labor market is initially in equilibrium and that then the government enacts legislation requiring a minimum wage rate above the market-clearing wage. Draw in this minimum wage on your diagram.

   a) How would you measure the level of unemployment on your diagram? Has it risen or fallen as a result of the minimum wage legislation?

   b) Do the relative magnitudes of the minimum wages and the unemployment rates in Quebec and Ontario support this result?

c) By how much does the level of employment change?

d) The legislation clearly raises the wage rate. Does it also raise total wage income? (Hint: Your answer will depend on a particular elasticity.)

e) What kind of people will become unemployed as a result of this legislation? Could discrimination of any kind play a role here?

f) Do you think politicians are enacting this legislation to deliberately hurt this sector of the economy? If not, what do you think is their reason for enacting this legislation?

2. The article states, in the first paragraph, that "not enough is known for sure about its impact on jobs, location costs and investment." What is your opinion of the impact of minimum wages on these three things? Explain.

3. What is the meaning of the statement that "Regrettably, less attention is paid to the competitive implications of a decision"?

4. The article notes that low-wage workers are typically found in tourism, retail trade and personal services (e.g., waitresses, car salesmen, and cab drivers). Can you think of any characteristics of these workers that might blunt the argument that they need minimum wages because their income is low?

I-8 B Rent Controls

Vancouver Sun, June 4, 1977

Removal of rent controls likely in some areas soon

Rent review commissioner Jim Patterson said Friday it would be reasonable for the provincial government to start removing rent controls in some sectors of the market and some regions where the market is operating well.

Patterson and rentalsman Barrie Clark agreed to discuss recent recommendations made to the provincial government for revisions in the Landlord and Tenant Act but both said they could not be specific because the matter is now before the cabinet.

Consumer Affairs Minister Rafe Mair has said he will introduce revisions to the act when the legislative session resumes June 13.

The minister said the present situation is slightly pro-tenant and a better balance will be struck.

Patterson said that, although Greater Vancouver now has quite a low vacancy rate, 1.6 per cent, there is a high rate in more expensive accommodation and the government could consider starting with control removals near the $500-a-month cutoff point.

Units which rent at more than $500 a month are exempt from the provincial rental increase ceiling of 10.6 per cent per year. (A bill now before the legislature would reduce the ceiling to seven per cent, retroactive to May 1.)

In addition, some areas have quite high vacancy rates and could be removed from the controls, Patterson said.

A vacancy rate of three to four per cent in any sector or region would allow consideration of removal of controls, he said.

Although the goal of the rent commission is to remove controls, Patterson said he objects to any inference that controls have caused a shortage in rental units.

"There is a very low correlation between housing production and rent controls and a very high correlation between housing production and other factors such as tax incentives and mortgage rates," Patterson said. "It really comes down to 'Can you provide affordable housing.'"

The rentalsmen said a number of housekeeping changes have been proposed to the provincial government but none involving a basic change in the philosophy of his office.

Clark said the rentalsman's office has proposed an approximate halving of the present time of 15 days before landlords can act against tenants who have not paid their rent. The 15-day appeal period on decisions of the rentalsman would also be approximately halved.

The rentalsman has asked for the power to subpoena and to order that the peace be kept.

Sometimes a landlord who is determined to remove a tenant will claim he must move his own family into the unit, Clark said, adding that he wants to be able to check out such situations and act to change them.

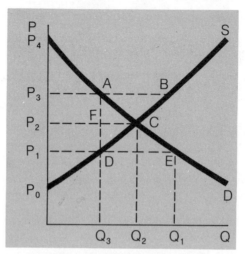

Figure I-8.2

1. Suppose Fig. I-8.2 represents the supply/demand situation for rental housing.

    a) Which price/quantity combination best represents the current situation with rent controls?

    b) Is there a shortage or surplus of housing? Identify it on the diagram.

    c) Would you expect discrimination (racial discrimination, for example) to increase or decrease when controls are enacted? Explain why.

    d) In terms of the diagram, how big is the consumer's surplus under controls? How big is the producer's surplus?

    e) If the controls were removed, what price and quantity would prevail? What would consumers' and producers' surpluses be? Is the sum of these surpluses greater than, equal to, or less than the comparable sum from question (d)?

    f) Does producer's surplus rise or fall when rent controls are imposed? Does consumer's surplus rise or fall?

    g) What does your answer to part (f) suggest about who gains and who loses from rent controls?

2. In the first paragraph Patterson talks of markets "operating well." What do you suppose is Patterson's definition of "operating well"? What would an economist's definition of this term be? Comment on the difference.

3. Explain why Patterson feels that "some areas have quite high vacancy rates and could be removed from the controls."

4. How can supply and demand analysis be used to explain the high vacancy rate in expensive accommodation but a low vacancy rate in cheaper accommodation?

5. What do you anticipate would happen to the price of non-rental housing if rent controls were imposed? If rent controls were eliminated? (Hint: Consider both demand and supply sides).

6. Comment on Patterson's statement that "There is a very low correlation between housing production and rent controls."

I-8 C The Oil Market

Toronto Globe and Mail, *February 26, 1976*

Institute proposes July jump in oil to world price level

By JAMES RUSK

The price of oil in Canada should go to world levels on July 1, the C. D. Howe Research Institute in Montreal says.

In its annual review of economic policy developments and the outlook for the future, the institute argues that this would be the best possible option for oil prices.

While a jump to world levels would be a shock to the economy, adding to inflation and curbing economic growth, "this approach has the advantage of being simple, direct and complete.

"Most important, it would bring into play quickly the incentives for conservation and development that are so badly needed."

A second-best solution would be to allow the price to rise to world levels over the next three years, with regular price increases of 67 cents a barrel at the beginning of each quarter, the report says.

A sudden sharp increase in the price of oil has two consequences that need discussion before a decision to proceed is taken.

A means would have to be found to offset the shock to the economy and the report suggests that the federal and provincial governments agree to sales tax cuts equivalent to the effects of the oil price increase on the consumer price index and personal disposable income.

Such a jump would also transfer a large amount of income, about $2-billion a year in the next few years, from oil users to governments and, to a lesser extent, oil producers.

"The urgency of conservation requires the user to pay the full amount, but what should governments do with the extra taxes and royalties on that higher price?

"So far, there has been remarkably little accountability by governments for the economic rent they have been collecting from producers and consumers."

Federal revenues have simply disappeared into general income, although some of the funds may have gone to investments such as Syncrude Canada Ltd. or Petro-Canada.

Alberta has formed a number of authorities to use the revenue but not all of these activities absorb the extra oil revenue.

Saskatchewan is considering using the revenue from oil to nationalize the potash industry.

"These rather hazy plans raise three questions:

"Are these the investments that make the best use of the funds that governments are collecting from the depletion of oil reserves?

"Does the consumer and taxpayer not deserve a more careful accounting for the money being earned on the sale of these national resources?

"Are governments using this temporary source of revenues (Canada will eventually run out of cheap oil reserves) as a means of financing new long range spending programs?"

The report also argues that the two-price system for oil in Canada is obsolete.

It was introduced at a time when it was believed that Canada would be reasonably self-sufficient in the future and the program would be self-financing.

"That rationale no longer exists.

"The system is not self-financing; Canada is a net importer, and the taxpayer is now paying a sizeable portion of the income subsidy.

"At the same time, the cost of lost efficiency has soared because of the net import position.

"It is no longer a matter of trading off revenues among consumers, producers and governments, it is a matter of how wide the gap between supply and demand can be allowed to go."

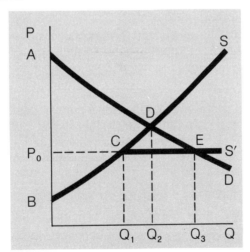

Figure I-8.3

1. If there were no limitations on imports or exports of oil, the supply curve for oil in Canada would be a horizontal line at the prevailing world price. (P_0S' in Fig. I-8.3.) With imports and exports banned, the supply curve for Canadian oil would be BS in Fig. I-8.3. With strict controls on exports and no controls on imports, the supply curve would be BCS'.

a) Is the sum of Canadian consumers' and producers' surpluses larger in the case in which imports are permitted or in the case in which imports are banned? Explain.

b) Would explicit recognition of the fact that Canada's maximum supply of oil (in the short run) is at a level less than Q_2 cause the difference in surpluses in part (a) to be greater or less? (Hint: Redraw the supply curve to reflect this).

c) How much oil would be imported if imports were permitted? How much would it cost to import it?

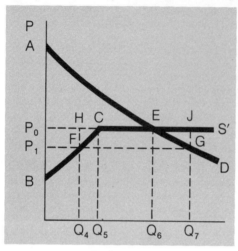

Figure I-8.4

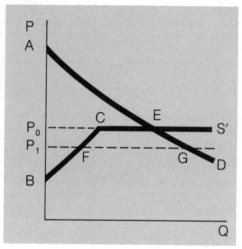

Figure I-8.5

2. Suppose that Fig. I-8.4 represents the Canadian supply/demand situation, but that the government is holding the price of oil at P_1, below the world price P_0, subsidizing imports.

   a) How much oil is supplied by domestic producers?

   b) How much oil is imported?

   c) How large is the subsidy paid to importers?

   d) What is the sum of consumers' and producers' surpluses for the case of no price control?

   e) What is the sum of consumers' and producers' surpluses for the case of price control? (Hint: Don't forget to subtract the government subsidy to importers—consumers will be taxed to pay this!)

   f) Which sum is larger?

3. Suppose Fig. I-8.5 represents the Canadian supply/demand oil situation with a fixed price P_1 set by the government below the market-clearing price P_0 (imports remaining subsidized). Now suppose the government taxes the domestic industry by imposing a per-barrel tax. (A careful reading of the article should reveal that this is their intention.)

   a) How would this action be reflected on the diagram?

   b) Does this lead to more, fewer, or the same quantity of imports?

   c) How can you represent the government tax revenues collected on your diagram?

   d) Does this action lead to a smaller or larger sum of consumers' and producers' surpluses?

4. What does the fourth paragraph of the article mean?

5. The article talks of collecting "economic rent" from producers and consumers. What does this mean? (Hint: Check a textbook for the definition of economic rent.)

I-9 Marketing Boards

An earlier section of this book (Section I-4) discussed the corn–hog cycle, in which markets are characterized by large swings in prices and output. This was due to producers, particularly in the agricultural sector of the economy, determining next period's output on the basis of this period's price. This led to gluts and shortages. Such instability in a market causes hardships for both producers and consumers. When a glut develops, prices drop and producers go bankrupt or suffer a severe loss of income; when shortages develop the price rises and consumers complain of being ripped off.

Although most economists favor free competition because of its efficiency, they usually admit that unstable competitive markets are sufficiently undesirable that government intervention in or control over such markets to prevent instability is needed. The loss in efficiency caused by the government intervention is more than offset by gains to society from having a steady supply of food (for example) at stable prices. Agricultural markets are renowned for their instability, for reasons discussed in Section I-3. Farmers tried to deal with this problem by forming co-operatives, but the voluntary nature of these organizations caused them to fail. The government has now set up marketing boards to deal with this problem.

Just because many economists admit that government intervention in or control over unstable markets is needed, however, does not necessarily mean that they support the methods actually used by the marketing boards to deal with this instability problem. In fact, many do not. Their basic complaint is that many marketing boards do not employ methods of stabilization that minimize the efficiency loss entailed by restricting the operation of free competition. Although this could result from stupidity or poor forecasting on the part of the marketing boards, it usually results from the adoption by the marketing boards of extra goals, beyond the stabilization goal. One popular goal is redistribution of income from consumers to farmers; another is maintenance of the "way of life" of small farmers. Attempts to meet these extra goals involve further distortion of the competitive market, further reducing economic efficiency.

Marketing boards do not all operate in the same way, since they must tailor their operations to the particular market with which they are dealing (the wheat market is more strongly influenced by international forces than is the egg market, for example), but they can in general be characterized as monopolies with the power to control supply and in some cases prices. Supply is controlled by preventing free entry to the industry through the use of output quotas for producers. (These quotas are usually determined, at the time of the marketing board's formation, by farmers' current output levels.) The quotas are valuable; as a market develops for the quotas, with the market-determined price of a quota right depending on the profits that can be made from producing that quota of output. (A quota right is the legal permission to produce a certain level of output.)

These profits in turn depend on the price of the output (often determined by the marketing board) and the efficiency of the producer. If the price is determined by the marketing board it is usually set at a level such that an average-sized farm will make a reasonable return. In a competitive market the smaller, inefficient farms will be forced out of the market and the remaining farms will expand to the optimal farm size; the price will consequently fall, forcing out of business all those farmers who do not move to the optimal farm size. With a marketing board, however, although the small inefficient farms will be forced out of business, there is little pressure on the less-than-optimal-size farms to increase to the optimal size (independent of whether or not the marketing board sets the price). This is because to increase his output a farmer must first buy a quota right. The money paid for a quota right could be invested elsewhere to earn a return approximately equal to the return from owning the extra quota right and producing the extra output. As a result there is little incentive for producers to expand to the optimal farm size; the industry will not be as efficient as it could be.

These inefficiencies usually manifest themselves in a higher-than-necessary price to consumers, and sometimes also appear in the form of surpluses of agricultural products. The disparity between the Canadian and the U.S. price for eggs, the selling of Canadian powdered milk on world markets at extraordinarily low prices, and the rotting of huge numbers of eggs in storage are examples of the costs associated with these marketing boards. Spectacular examples such as these inevitably lead to newspaper comment.

I-9 A Fresh Eggs

Vancouver Sun, May 30, 1977

'Fresh' eggs often aren't

By EVAN EVANS-ATKINSON
Sun Consumer Reporter

PARKSVILLE—B.C. consumers at times are sold Grade A eggs that are up to three months old, the B.C. food inquiry was told today.

Sidney egg farmer W. T. Burrows blamed the delay on policies of the national and B.C. egg marketing boards.

"Because of the wide latitude of quality allowed in Grade A eggs...the consumer, depending on location and time of year, will receive considerably less quality for his money depending on where he buys his eggs," Burrows said in his brief to the food inquiry committee.

"The time interval between egg lay and purchase by consumer can range from three days or less to three months with eggs still meeting the Grade A requirement."

Burrows' brief was among the first presented to the 10-member legislative committee as it opened two days of public hearings here.

Subsequent hearings will be held in Vancouver and throughout B.C. as part of its year-long probe into the high cost of food in B.C.

Burrows said the Canadian Egg Marketing Agency causes much of the delay in selling eggs because it takes the estimated demand for eggs for one year and divides it evenly over 12 months.

"The result is a uniform monthly supply which fails to take into consideration seasonal variation in demand.

"Therefore the grading stations find that at some times of the year it is necessary to store eggs because they are in excess of demand and they can anticipate an increased demand a month or two away.

"When the anticipated time arrives, the stored eggs begin to move out—the oldest first—until the stock is current.

"This 'CEMA defect' covers the whole province," said Burrows who has a master's degree in agriculture.

The provincial egg marketing board's uneven distribution of egg production quotas to farmers further contributes to delays in selling eggs, he said.

Burrows cited the example of Fraser Valley eggs having to be shipped to Prince George because egg producers in the north do not have enough quotas to fill local demands.

He said the value of quotas in the past few years has risen more than 400 per cent and the result is that only large farms are able now to borrow money to buy them.

The small egg farmer in B.C. is either being pushed out or bought out of business because of this, he added.

Burrows said he has been told the large feed companies are providing loans to buy quotas which leads to farmers having to buy feed from the companies which loaned the money rather than from smaller companies offering lower feed prices.

Higher prices for feed and the interest on loans result in increased cost that are one way or another passed on to the consumer.

"It is doubtful if there is more than token competition on feed prices anymore. It is evident they are competing to obtain quotas for their customers."

Burrows emphasized that a well-run marketing board is essential for a sound program of production, distribution and fair prices to all.

"The quotas system is the disease which will bring about the death of marketing boards if allowed to go on, he said.

"The consumers will be the winners in the battle which they are now losing."

1. In the absence of the marketing board, how would the problem of variability in the freshness of eggs, as described in the article, be solved?

2. Why doesn't the egg marketing board solve the problem of having to ship eggs from the Fraser Valley to Prince George by simply allocating more quotas to farmers in the Prince George area? How would you suggest they solve this problem?

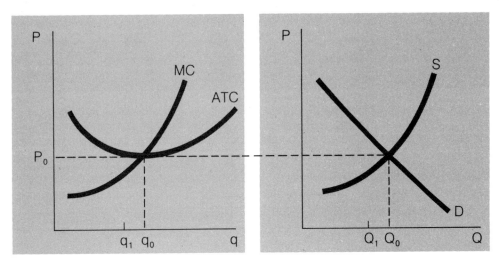

Figure I-9.1

3. In Figure I-9.1 suppose that before the marketing board is created the egg market is characterized by the typical farm producing quantity q_0 at a price P_0, with a total supply of Q_0. Now suppose that the marketing board introduces a quota system, permitting the typical farmer to produce only q_1, moving the total output from Q_0 to Q_1.

a) What happens to the market price?

b) What happens to the typical farmer's income? Quantify your answer by using the diagram.

c) How would the value of the typical farmer's quota be determined?

d) If the marketing board had used the quota system for stabilization purposes only (i.e., rather than also using it to raise farmers' incomes) what quota level would they have set? How would this affect the answers to parts (a), (b) and (c) above?

e) What would happen if the quota level set in part (d) were kept constant over time? What changes in quotas would be required?

f) Does the rise in income generated by the quota system apply to new as well as old farmers? Explain why or why not. If possible, interpret this on the diagram.

Some of the best scraps are over food

Financial Post, July 31, 1976

By Sheldon E. Gordon

OTTAWA — HOW YA' gonna keep 'em down on the farm after they've seen what's in the national food policy recommendations drafted by federal government officials?

The secret document — approved for cabinet consideration by an inner circle of 10 deputy ministers — was bared in a Toronto Star article last week. Its contents provoked more commotion than the proverbial fox let loose among the chickens.

Just two months ago, the Canadian Federation of Agriculture had been assured by Prime Minister Pierre Trudeau that Ottawa was merely reviewing its existing farm programs: no dramatic changes were contemplated.

Meanwhile, though, the Privy Council Office was co-ordinating the drafting of policies which would drastically alter the conditions under which Canadian farmers produce and market their foodstuffs.

The proposals embody the philosophy of the defunct Food Prices Review Board. Its former research director, Al Lyons, has become the leading in-house proponent of a "national food policy" in his current position as assistant deputy minister of Consumer & Corporate Affairs.

But cabinet adoption of the draft policy is far from assured. Agriculture Minister Eugene Whelan reacted angrily to the document, dismissing the opinions of the deputy ministers who approved it — they did not include his own DM — as out-of-touch with agricultural realities.

Even Consumer Affairs Minister Bryce Mackasey was not overly impressed after reading the document. Its destiny should become clearer after it is discussed in a cabinet committee meeting scheduled for next week.

In essence, the draft policy would assign a greater role to market forces, and a lesser role to government and to producer-run boards in determining levels of farm output and returns to the producer.

In recent years, federal agriculture policy has been as volatile in its ideological swings as the price fluctuations in the agricultural markets it was designed to stabilize.

On the one hand, Wheat Board Minister Otto Lang has introduced measures with a market orientation. Interprovincial marketing of feedgrains, long the preserve of the Wheat Board, was opened to private traders. A commission was created to study removal of Ottawa's Crow's Nest subsidy on the rail movement of Prairie grain.

Meanwhile, Whelan has pressed contrary measures. National marketing plans were launched to control production levels and prices for commodities such as eggs and turkeys. Where Ottawa did not give farmers the right to set prices outright, it intervened with "support payments" to dairymen, cow-calf operators, and potato growers. Imports of eggs, cheese and beef have also been restricted.

The top government bureaucrats now apparently want some consistency in food policy. They want supply management schemes and price supports dropped in order to remove distortions in the marketplace. They want freer entry for some food imports. They want, in short, consumers' interests to be protected by more competition in agriculture.

The CFA suspects that what those Ottawa city-slickers really favor is a return to the bad old days of the "cheap food policy."

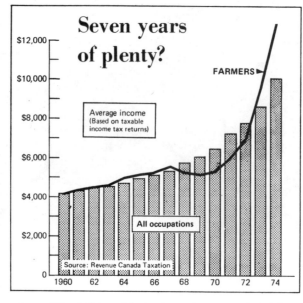

Seven years of plenty?

Average income (Based on taxable income tax returns)

FARMERS

All occupations

Source: Revenue Canada Taxation

1960 62 64 66 68 70 72 74

Between 1960 and 1970, retail food prices rose by 32%, virtually the same increase as all other retail prices. But the prices actually received by farmers during the same period climbed by only 21%.

The average income of farmers who earned enough to pay income tax was either roughly even with — or lagged behind — that of the average for all occupational groups during the same period. As late as 1972, the farmers' average was $6,954 (vs $7,784 for all occupations).

Only in recent years has the trend changed. In the 1971-75 period, while prices of other retail items climbed by 30.5%, retail food skyrocketed by 62%, and the prices received by producers jumped by 95%.

By 1974, average income of farmers surpassed that of the average for all occupational groups ($12,932 vs $10,038).

This remarkable improvement, according to farm spokesmen, really reflects the high world grain prices of the past few years. They say that the average income figure gives a misleadingly rosy impression of other commodity producers.

They also hasten to add that Ottawa has acted to cushion Canadian consumers from the full impact of the grain boom. A two-price system was introduced to make Canadian wheat less expensive here than abroad. (But steeper world feedgrain prices, which were not controlled domestically, boosted the production costs of Canadian poultry and livestock.)

The CFA, in its May submission to the cabinet, lamented the public's "exaggerated preoccupation with the alleged exploitative potential of the so-called monopoly powers of marketing boards."

In Ontario, only seven of the 20 marketing boards set prices for their commodities, and only two — eggs and fluecured tobacco — set production levels. If the price of eggs set by the Canadian Egg Marketing Agency (Cema) were lowered by 5¢ a dozen, the average family of four would save only $4 a year, the CFA claimed.

"The instability of the poultry and egg cycles... was intolerable. The end result would inevitably have been a very heavy degree of concentration in the industry, already sufficiently great, with little or no price benefit to consumers over time..."

It's a seductive argument. Supply management schemes and price support payments foster competition by ensuring that severe price fluctuations don't ruin smaller operators, leaving the field to the big producers.

But perhaps the reasoning is flawed. University of Alberta rural economists Terry and Michele Veeman contend that marketing boards and stabilization payments only aggravate the inequitable distribution of farm income.

Boards which regulate output usually assign quotas on the basis of past production levels. This reinforces the position of the largest producers, and inhibits structural change in the sector.

"In addition," the Veemans noted in a recent presentation to the Agricultural Institute of Canada Conference, "problems arise from the tendency for program benefits to become capitalized into the value of quotas."

Support price payments help the farmer at the expense of the taxpayer rather than of the consumer, and are therefore more visible. But they also aggravate income inequities.

"The federal dairy program is a classic example," the Veemans say. "Since benefits are essentially distributed on the basis of sales, this program does relatively little to enhance the income position of the smallest and marginal producers.

"Nor does it offer any encouragement toward reallocation of their resources into more productive areas."

By the end of next March, the Canadian Dairy Commission will have paid out $750 million in subsidies to Canadian dairymen over a 3-year period. Although not sufficient to keep them from showering Eugene Whelan with skim milk, the total is a whopping 128% increase over subsidies paid in the previous 3-year period.

Ottawa, under the Agriculture Stabilization Act passed in 1958, has paid out $2.06 billion, mainly to dairymen, cow-calf operators, and potato and sugar beet growers.

The act was amended last summer to permit support price payments for nine "named" commodities and others which might be added by order-in-council.

The changes guarantee producers a per unit return equal to 90% of the average price of the commodity at representative markets over the previous five years. The formula is indexed to take account of rising production costs.

Some of the individual commodities, such as industrial milk, have cutoff points: the subsidy is paid only up to a certain volume of output per producer. But whereas the U.S. has a $20,000 ceiling on the total value of subsidies for which a farm unit is eligible, Canada has no such overall maximum.

Despite criticisms that the bulk of stabilization payments goes to farmers who do not need the money, Agriculture Canada insists it has no statistical profile to indicate the distribution of subsidies among upper- and lower-income farmers.

There is also doubt that "stabilization" payments actually stabilize. The Veemans maintain that the various price support subsidies dispensed by Ottawa between 1939-1974 boosted net farm income 6% above what it would otherwise have been. But the payments "did relatively little to stabilize year-to-year changes in net farm income."

Supply management and stabilization payments, they suggest, are the wrong ways for government to aid the farmer. These measures impede structural changes in farming by protecting specific commodities and existing producers.

Moreover, raising the *average* level of prices and incomes is an inefficient way of helping low-income farm families: most of the benefits go to big, profitable farm operations.

Instead, the Veemans urge, the needs of poor producers should be met directly — through some form of guaranteed annual income for farm operators.

This more selective approach is bound to raise objections. Farm groups, predictably, will argue that average returns must be kept above the level of market prices if there is to be enough incentive for producers to go all out, rather than to curb output to raise prices.

Humanitarian souls, listening to the rumble of empty stomachs in the third world, may be inclined to agree. But do we really want to encourage Canadian farmers to expand production through incentives?

The results could be a Lift program, in reverse. That program paid grain producers to take land out of production in the early 1970s and helped deplete the world's granaries to dangerous levels. Paying Canadian farmers to grow more might well mean overproduction (which already afflicts the dairy and cattle sectors) or inefficient production (if the sugar beet industry were revived).

Domestic demand for food is expected to grow less rapidly, rather than more rapidly, in future years. Foreign demand would be unable to gobble up the Canadian surpluses — because the poorer nations, though needing larger food volumes for their undernourished people, lack the purchasing power to support it.

Incentives to produce? Support prices have acted as an incentive to overproduce. Marketing boards try to avoid that problem by aligning supply with demand.

The early experience of Cema, however, shows that they don't always succeed. Too many yolks is a heavy yoke to bear.

1. To what factors is Whelan alluding when he says the new draft policy is "out-of-touch with agricultural realities"?

2. It is suggested that the "draft policy would assign a greater role to market forces." What does this mean?

3. Why might Lang take a more market-oriented approach with the Wheat Board than Whelan has with other agricultural products?

4. The article compares the average income for farmers with other occupations. In what ways could a statistician bias this comparison so as to make the farmer appear better off?

5. To what is the CFA referring when they say the "instability of the poultry and egg cycles . . . was intolerable"?

6. What is implied about the long-run average cost curves of the poultry and egg industries which would create a "very heavy degree of concentration"? What does the statement "with little or no price benefit to consumers over time" imply about these cost curves? Are these two statements compatible? Explain why or why not.

7. Evaluate the "seductive" argument that marketing board schemes foster competition.

8. What does it mean to say that the program benefits "become capitalized into the value of quotas"? What problems arise from this tendency?

9. Indicate the direction of change of quota values as a result of each of the following changes. Briefly explain your reasoning.

   a) elimination of barriers to imports;

   b) increase in the domestic quota;

   c) a ceiling on the total subsidy collected by a farm unit;

   d) a subsidy to small farmers only;

   e) an increase in demand.

10. The Veemans suggest that poor producers should be given some form of guaranteed annual income. How would this avoid the problems that they state are associated with supply management and stabilization payments?

11. What is the meaning of the last sentence in the article?

12. The article discusses two basic policies, "supply management," in which output quotas are imposed, and "price supports," in which the price to the farmer is held above the market price. Draw a supply-demand diagram for each of these cases, indicating the price and output that would result. Comment explicitly on the role played in each of these cases by imports and exports. Discuss the relative merits of the two policies. (Be sure to consider alternative ways of getting rid of a surplus.)

I-10 Foreign Competition

Because Canada is a small, open economy, its markets are kept competitive mainly by foreign competition. Although this ensures that Canadians are able to buy from the world's lowest-cost producers, it also means that domestic industries are limited to those in which Canada has a definite comparative advantage. Not wanting to be a nation of "hewers of wood and drawers of water," Canada has encouraged development of manufacturing industries, often protecting them with tariffs or quotas.

This protection, in the face of economists' advocacy of free trade, has been rationalized in several ways. First is the "infant industry" argument: To get an industry started and functioning efficiently requires a few years of protection from competition; once on its feet the protective trade barriers can be removed. Second is the "vital industry" argument: Although it may be cheaper to buy a good elsewhere, in the event of a war or some international emergency, an assured supply of that good may be thought necessary. Third is the "nationalism" and "prestige" argument: All developed countries, some people think, should have certain industries operating in their economy either for "prestige" or to enhance their "nationalism." Fourth is the "uncertainty" argument: Foreign suppliers may be unreliable, implying large fluctuations in quantities delivered and prices charged, disrupting domestic industry using this good as an input. Fifth is the "foreign monopoly" argument: It may be the case that foreign suppliers may act as monopolists (or as a cartel) in the absence of domestic competition, forcing Canadians to pay a monopoly premium. Sixth is the "diverse economy" argument: Canadians may perceive distinct social advantages (e.g., giving citizens a wider choice of occupations) and stability advantages to having an economy characterized by a wide variety of industries, rather than being reliant on the fortunes of and opportunities provided by a small number of industries.

In response to these arguments, the Canadian government has imposed some tariffs and quotas to protect certain Canadian industries. At the same time, however, the Canadian government recognizes the benefits of trade, and in negotiations with other trading countries is constantly pressing for lower tariffs imposed by others and is in turn being pressured to lower her tariffs. Lowering Canadian tariffs accomplishes several objectives. First, it serves to increase trade and the benefits derived therefrom. Second, it keeps domestic industries competitive, ensuring fair prices for Canadian consumers and acting as an anti-inflation force. And third, it acts as a negotiating instrument in obtaining reductions in foreign tariffs, aiding Canadian exporters.

Thus although the Canadian government protects Canadian industries for the reasons given earlier, it keeps that protection at the lowest level it thinks is feasible. As a result, industries are constantly complaining that their protection is insufficient. In the 1970s this was aggravated by the relatively large wage increases granted Canadians and the low investment in capital goods caused by the uncertainty surrounding wage/price control programs.

I-10 A **Double-Knit Bankruptcies** Toronto *Globe and Mail*, April 10, 1976

Double-knit bankruptcies linked to imports

By ALBERT SIGURDSON

More than 20 Canadian double-knit companies have gone out of business in the past year mainly because of low-cost imports, according to William Berry, economics director of the Canadian Textiles Institute, an industry-sponsored body. It is a crisis only Ottawa can solve, says the institute.

And Ottawa's Textile and Clothing Board has now been pressed into holding a review on double-knit and warp-knit fabrics that will start early next month.

Canadian textile producers are being hit indirectly through massive and increasing imports of garments. Double-knit apparel imports from Taiwan alone more than doubled in 1975 to 11.7 million units, from 5.7 million a year earlier.

This means that Canadian apparel makers are switching out of manufacturing and are becoming importers, further reducing the base market for Canadian textile producers.

Then, in order to compete, the Canadian producers are forced to price their product so low that they cannot make money. Close to 500 double-knit machines have gone out of production in the past year, Mr. Berry said in an interview.

About 22 million pounds of double-knit fabric from all sources were imported last year—the bulk of the Canadian requirement—and it appears the Canadian manufacturers are being driven out of their own market, he said.

The problem was allowed to get so severe because the textiles industry is not permitted to seek remedy to imports of garments, only to imports of fabric.

"There's a lot of sales (of imports) below cost going on in this country. You can get remedy through the Anti-Dumping Act, but that takes about a year, and we haven't got a year," Mr. Berry said.

As a result, the brief will ask the Textile and Clothing Board to halt the flow of imports of apparel and a surtax on those that do come in.

It probably will ask for the maintenance of existing restraints on the import of fabric from Taiwan, South Korea, Hong Kong and Japan and the introduction of restraints on any other low-cost supplier.

The Textile and Clothing Board is currently reviewing the market for worsted fabrics, used mainly for men's apparel, and producers' complaints about a high level of imports.

"We've got to get our share of that men's market up to 80 per cent," Mr. Berry said, in order to justify investment in new shuttleless high-speed looms.

"We desperately need new machinery," to reduce unit labor costs and improve quality. Quality is good now, but although the Canadian market is relatively small, Canadian manufacturers have to compete in a sophisticated demand setup, Mr. Berry said.

"Where the Chinese run one brown gabardine, the Americans might run three—but we've got to run six."

The Canadian producers react well and quickly to the needs of domestic apparel makers, but this means small-volume runs and higher costs.

They have about 60 per cent of the apparent worsted market now, but foreign producers have had price leadership for years.

Canada imports worsted fabrics from 28 countries, but only two are under any form of restraint, Mr. Berry said.

"We can't even tolerate the 6 per cent annual increases in the two remaining restraint agreements that we have." He said in order to get 80 per cent of the market as intended in five years, domestic producers would have to get all of the growth.

The federal Government could move to hold imports at the current levels because the General Agreement on Tariffs and Trade provides for extraordinary restraints if the importing country's domestic industry is threatened.

The competitive pricing situation is complex, but "the men's wear industry needs us. Foreign prices are not stable. When foreign prices go 'way up, we tend to keep ours down. We give service at relatively shorter lead times."

Mr. Berry said Canadian worsted mills are large by world standards—"right around optimum size." With new machinery they could better meet price competition from countries where wages are lower.

---

1. The Law of Comparative Advantage states that to maximize welfare, countries should specialize in producing those goods in which they have a comparative advantage. It appears from this article that Canada might not have a comparative advantage in producing double-knits or worsteds and thus would be "better off" if the resources in this industry were applied to the production of some other good in which Canada does have a comparative advantage. Several arguments were given in the introduction to justify the application of trade barriers. Which of these arguments are alluded to in the article itself? Be specific in your answers.

2. What is the Anti-Dumping Act? What is its purpose? Why would anyone want to violate this Act?

3. What statement in the article supports an economic analysis of the double-knit industry based on the assumption of an internationally-determined price?

4. The fact that twenty Canadian double-knit companies have gone out of business indicates that the internationally-determined price is below which of the following: (a) marginal cost; (b) average total cost; (c) total cost; or (d) average variable cost? Explain your answer.

5. Explain the meaning of the statement "Where the Chinese run one brown gabardine, the Americans might run three—but we've got to run six." Of what relevance is it to the cost function of Canadian firms? Does this mean that Canadians are unable to take advantage of economies of scale? If so, how do you reconcile this with the statement that Canadian worsted mills are large by world standards—"right around optimum size"?

Microeconomics 43

6. Why would the share of the men's market for worsteds have to be at least 80 percent to justify investment in new shuttle-less high-speed looms? Explain your answer by using a cost-curve diagram.

7. The Canadian Textiles Institute intends to ask the Textile and Clothing Board to "halt the flow of imports of apparel and a surtax on those that do come in." Let us look at these two options, import quotas and import tariffs. Suppose the Canadian demand and supply of double-knit apparel is given in Fig. I-10.1.

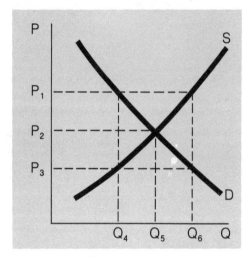

Figure I-10.1

a) What combination of price and output do you feel best represents the current situation? Explain.

b) What is the current level of imports, as shown in the diagram?

c) What does the demand curve faced by importers look like? Refer specifically to the diagram in explaining how it would be derived.

d) Suppose a quota is imposed on imports, equal to half of the current level of imports. Draw a new diagram showing the new price and output levels. Explain how you obtained your answer.

e) When a quota is imposed there will be "quota profits." What are quota profits and to whom do they accrue? Explain how they would be measured in your diagram from part (d). Can you think of a way by which the government could reap these quota profits for itself?

f) Suppose an ad valorem import duty (tariff) were imposed. (Such a tariff consists of a percentage of the value of the import. An ad valorem tax of 30% means that if the imports originally sold for $5, a tax of $1.50 would be payable.) Draw a diagram showing the new domestic price and output levels.

g) Explain how the duty collected would be measured in your diagram in part (f).

h) Under which system (quotas or tariffs) would the domestic producers benefit most (or lose least) in the event of

    i) an exogenous increase in demand?

    ii) an exogenous decrease in demand?

    iii) an exogenous increase in the world price?

    iv) an exogenous decrease in the world price?

i) What are the relative merits of the quota and import tariff schemes? Take both the government's (and consumers') and the producers' point of view.

I-11 Imperfect Competition

In perfect competition each seller faces a perfectly elastic demand curve. Imperfect competition is characterized by sellers facing downward-sloping demand curves, either because they are monopolists and face the industry demand curve, or because their competition produces a similar but not precisely identical (at least in the eyes of the consumer) product. In the latter case economists talk of an oligopoly when the number of firms is quite small, and monopolistic competition when the number of firms is large (and all are about the same size).

Oligopolists play a cat-and-mouse game with their fellow oligopolists. Because there are only a few firms, the actions of any one firm have a significant effect on the profits of the other firms. Any action by one firm causes other firms to change their behavior accordingly, affecting everyone's profitability. Because of this firms try to second-guess their competition, incorporating in their decision-making procedures a guess at how their rivals will react. Since there is a wide variety of different kinds of guesses an oligopolist could make about his rival's reactions, there are several different "theories" of oligopoly behavior, no one of which has proven to be "best."

Monopolistic competition, on the other hand, does not suffer from this dilemma. The large number of competitors, combined with freedom of entry, removes this problem and allows, as your textbook should testify, a standard theory explaining monopolistic competition. The blue jeans market, discussed in the first article below, is not perfectly competitive, since jeans differ in quality and fit as well as brand name. Although this market is not characterized by a large number of sellers all of roughly the same size (there are a few big companies along with smaller ones) the monopolistic competition model can be usefully applied to analyse the contents of this article.

The second and third articles below relate to the automobile market, an oligopolistic market. Although it would be appropriate for us to ask questions relating to oligopoly theory about these articles, we have chosen not to do so. This is because the wide variety of existing oligopoly theories precludes a definitive economic interpretation of oligopoly behavior, forcing questions related to oligopolies to be somewhat nebulous. Instead we have chosen to use the story reported in these articles as a vehicle to discuss a phenomenon (accentuated in oligopolistic markets) that has important implications in both microeconomics and macroeconomics. This is the distinction between price and quantity adjustment as a response to disequilibrium.

Whenever a market is in disequilibrium, supply does not equal demand; the "right" price/quantity combination has not been found. In general both price and quantity change in response to a disequilibrium, but usually one changes faster than the other. In some markets price changes are faster than quantity changes; let us call this a price-adjusting market. In other markets changes in quantity are fast relative to changes in price; let us call this a quantity-adjusting market. The markets described in the articles in the first two sections of this book are characterized by price-adjusting dynamics: price responds to disequilibrium more rapidly than does quantity.

The automobile market is a quantity-adjusting market in which quantity adjustments are faster than price adjustments.

Although economists have been aware of the existence of quantity-adjusting markets for some time, the importance of this different type of dynamic reaction was not recognized until the late 1960s when this theory was used to describe and explain Keynesian theories of prolonged unemployment. When faced with a fall in demand, businessmen are cautious. In general, they have no way of knowing (in the short run) whether the decrease in demand is permanent or temporary; as a result they react slowly to the disequilibrium situation. Instead of cutting price immediately, which could be costly (administratively as well as revenue-wise) they instead cut output, creating unemployment; in the short run they are quantity adjusters rather than price adjusters. The result is that unemployment is both more severe and more prolonged.

(A more complete discussion of the quantity-adjuster phenomenon in the context of unemployment would include a discussion of the labor market, in which labor is also postulated to be a short-run quantity adjuster rather than a price (wage) adjuster. This is beyond the scope of this introduction, however, and unnecessary for the understanding of the distinction between price and quantity adjusters.)

This quantity-adjusting phenomenon is particularly characteristic of oligopolistic markets, in which the participants are reluctant to change prices for fear of precipitating a costly (for them) price war.

'No, no, don't leave us to Levi'

By JUDY LINDSAY
Sun Business Writer

Among the fashion plates who showed up at the Vancouver hearings this week of the Textile and Clothing Board was a rangy young man whose long legs were encased in ordinary blue jeans.

He placed four pairs of blue jeans on the table in front of board chairman Gordon Bennett. The quality and fit were almost identical, he said, but one pair of jeans sells for $10 while the others go for $24.95.

He then proceeded to make a case for dropping the textile import quotas imposed by the federal government last November to protect Canadian garment-makers. Importers will be allowed to bring in only 90 per cent of their 1975 volume.

His intention, he said, was to show how some major Canadian garment companies push up the cost of their goods to Canadian buyers, and how the government with its quotas has become an accomplice.

His name is Wayne Overland, 36, and he is the major owner of six Jean Joint stores in Alberta. His philosophy, he told the board, has been to seek a higher volume of sales although it may mean smaller profit margins.

"This philosophy is apparently contrary to the policy of some of my suppliers who prefer to sell fewer units with a greater markup. They limit the supply and increase the prices.

"The Jean Joints' two best-selling brand name jeans are Levi and Howick. Both companies keep retailers on a quota system. You get only about half the units you could sell in key styles," Overland said.

He said pressure put on the manufacturers by other retailers who complained about his discount prices forced him sell the jeans at the regular retail price.

As the big companies pushed up their prices until blue jeans were selling for more than $20, Overland said he determined to offer a $10 jean again. A sports reporter for the Edmonton Journal, he went to Hong Kong on his holidays and there he said he discovered it was possible to make high quality blue jeans and cords that would land in Canada at half the price of brand names.

This jean he compared with the best-selling Levi. "The denim is of the same weight, both are 100-per-cent indigo (mine has been tested in the Eaton labs for shrinkage, etc.), the quality of construction and fit is if anything better in my model.

"There are two big differences. One doesn't have the Levi label. Mine lands me $6 in Alberta while Levi are charging me $12.95. I am selling mine for $10 and the Levis for the suggested retail of $24.95."

Levi both imports jeans and manufactures them in Canada.

Overland handed in his press card so he could devote full time to jeans. He found that some of the companies he dealt with in Hong Kong also supplied the Big Three—Levi, Lee and Wrangler—and he was able to get a good idea of their fabric and production costs.

"It became clear because of their volume they were negotiating cheaper prices than a small retailer like myself. Then I would see a company like Lee charging me $10.50 for jeans made in Hong Kong and Levi $10 for a cheaper model when I knew their true landed cost was less than $6.

"Instead of using these cheaper imports to average down the price of Canadian production, the prices are inflated, more proof of my contention the policy of these companies is to gouge the consumer."

Overland said he is not concerned about the jobs of Canadians at domestic garment factories; all he cares about is the jean-buying consumer.

The largest-selling jean in the world, he said, is the Levi unwashed bell bottom. Made in Hong Kong, the jean retails for $15 in the U.S. and $19.95 in Canada. The import duty to the U.S. is 16 per cent, to Canada, 22.5 per cent, a difference of 6.5 per cent, he said. (That's 6.5 *percentage points*; the Canadian duty is actually nearly 50 per cent higher than the U.S. duty. Nevertheless, said Overland, it still doesn't justify the amount of the Canadian markup.)

"Then why is the price difference more than 30 per cent? It is little wonder that Fortune Magazine lists Levi Strauss as the most profitable garment company in the world."

But Overland continues to sell Levis. "Because the public is so brainwashed by their heavy advertising they'll pay more than twice the amount for the same product because of the brand name. At the same time the success of my $10 jeans indicates many consumers recognize a better deal."

A believer in "pure free enterprise," Overland said he does not advocate the government investigate the way these companies operate, but rather it should allow small companies to compete and thus promote an "honest" market.

"Companies like mine that can offer the consumer high quality goods at low prices will do more to prevent unfair pricing than any laws.

"There is no need for quotas in the Canadian jean business. Most domestic companies are prospering. The only major case of failure I know of is Monarch Wear in Winnipeg. I can offer the board first-hand knowledge of how this company got into trouble.

"It was a case of mismanagement. For five years Monarch Wear was the main supplier to the Jean Joint. One of the owners became a personal friend.

"After the company went public it expanded too fast in all directions (much of it importing), management made bad decisions and lost control.

"As an example, in one back-to-school period, which is the key time of year in this business, Monarch Wear did not have a satisfactory supply of heavy denim. This is like having an airline without planes. The goods they shipped were grossly inferior and different from the samples shown. Is it any wonder we looked for other suppliers?"

"Far more importers have gone broke than domestic producers. The reason for this is offshore production requires endless supervision to ensure quality. As an example, I flew to Hong Kong eight times in the last 11 months as well as having an agent on the job fulltime to supervise production."

Companies should be allowed to appeal for increased quotas on the grounds that their goods are sold at prices beneficial to the consumer, he said.

Levi has huge quotas for 1975 "on which they make exorbitant profits," contended Overland. "If there is a quota there will be no new importers like myself to force the prices down. This plays right into Levi's hands."

1. Draw a diagram representing a monopolistic competitor and use this diagram to explain the statement in the sixth paragraph.

2. What statement in the article indicates that Levis are strongly differentiated from other jeans? What impact does a more strongly differentiated product have on your diagram in question (1) above?

3. How does Overland's move to arrange his own supply of jeans from Hong Kong fit into the textbook theory of monopolistic competition?

4. What effect does the government's imposition of import quotas have on the adjustment mechanism described in question (3) above? Does this support the accusation in the fourth paragraph that the government is an "accomplice"?

5. Interpret the final paragraph in terms of the theory of monopolistic competition.

6. Why would the price of Levis be so much higher in Canada than in the United States? (Hint: Consider Canada and the United States as separate markets with either different demand curves or different degrees of competition.)

7. Comment on Overland's general philosophy, with particular reference to the jean-buying consumer, Canadian garment-makers, unfair pricing, unemployment, and government quotas.

I-11 **B Car Glut** Toronto *Globe and Mail*, January 6, 1976

Car glut but no cut

Big Four in U.S. have 1,700,000 unsold autos, but say they cannot lower prices

© New York Times Service

DETROIT—When will the automobile industry cut prices?

Many dealers, consumers, analysts and economists are wondering why Detroit's top executives, who are vociferous in their defence of the free-enterprise system, do not respond to one of the system's basic laws—when you have a glut of goods, you cut prices to move them.

The four automakers here have a glut of about 1,700,000 unsold cars.

But when industry executives are confronted with the price question they say only that they are carefully weighing the balance between prices and production costs. At this point they say that they are going to hold prices where they are.

The automakers assert that the huge price increases on 1974 and 1975 models were, in fact, not large enough to offset cost increases. The ballooning of the costs of basic raw materials—steel, copper, zinc, aluminum—after the lifting of price controls was the main factor in increasing production costs. These materials costs have begun to decline in the past few months, but, the easing has been small and is just now beginning to flow through to the assembly plants where it might affect production costs.

From the time when controls were first imposed in 1971, auto executives argue, the industry has had to absorb a considerable part of its cost increases. General Motors estimates that since 1971 it has not recovered about $500 in the cost of producing the average car and truck.

Profit decline

To back their case the top executives point out that despite the price increases, the net income as a per cent of sales—the industry's profit margin—has declined dramatically. During the good years the ratio averaged about 6 to 8 per cent. But in the first nine months of 1974 it stook at 3.1 per cent for GM. 1.7 per cent for Ford, and eight-tenths of 1 per cent for Chrysler.

"You can't reduce prices very much and keep your business together on that basis." said Richard Gerstenberg, the former chairman of General Motors.

But wouldn't a price cut stimulate sales and thus increase the volume of production and reduce the cost per car? The industry holds that such a course now would be too much of a gamble.

Management has finally conceded that the high price of cars has been a major factor in preventing people from buying. But they insist that consumer uncertainty about the economy is an even greater factor.

Higher price for food and other basic items, layoffs, high interest rates and difficulty in getting credit have made many consumers very reluctant to make big purchases such as cars.

A price decrease in such an uncertain atmosphere, top executives believe, would have some effect—but not a great one—on car sales.

Must act first

They argue that President Gerald Ford and Congress must act first to cut taxes, cut interest rates and in general stimulate spending.

Some in the industry, particularly at General Motors which is the price leader in the industry, are apparently just beginning to see small signs of encouragement.

Thomas Murphy, the new GM chairman, observed in his year-end statement that interest rates have fallen and that inflation has moderated and that by next year productivity should increase and the real income of workers should grow.

"Our outlook for General Motors in 1975 anticipates a gradual sales improvement as the year advances," he said.

So for the present the strategy of the industry is not to reduce prices but to ride out the sales slump by cutting costs and production. In January, for example, about 300,000 out of the industry's 700,000 blue-collar workers and another 25,000 white-collar workers will be on indefinite and temporary layoffs.

All four producers are also trimming costs, re-evaluating future products and phasing out older plants. As one top auto executive put it, "We want to make sure that for every dollar spent we get a dollar back."

The hope is that when there is an upturn in sales the auto companies will have trimmed costs enough while keeping prices up, to ensure them a higher rate of profitability.

© 1976 by The New York Times Company. Reprinted by permission.

---

1. Draw a supply and a demand curve for automobiles and use this diagram to characterize the situation in the auto industry at the time this article was written. How would the situation before the glut appeared be characterized by your diagram (i.e., what probably happened, in terms of the diagram, to cause the glut)? How would a cut in prices solve this problem, in terms of your diagram?

2. The article cites "one of the system's basic laws—when you have a glut of goods, you cut prices to move them." Consider an alternative "basic law": "When you have a glut of goods, slow down or stop your production of them." Call the first "law" the "price-adjust" law and the second law the "quantity-adjust" law. Discuss the advantages and disadvantages, from a producer's point of view, of following the price-adjust versus the quantity-adjust law.

3. Discuss the advantages and disadvantages from society's point of view of having the producers follow the price-adjust versus the quantity-adjust law.

4. For the statement "But wouldn't a price cut stimulate sales and thus increase the volume of production and reduce the cost per car?" to be true, what must be assumed about (a) excess labor supply before the glut; (b) economies of scale; and (c) demand elasticity?

5. Explain the effect on the supply/demand diagram of the following: consumer uncertainty, higher prices for food and other basic items, layoffs, high interest rates, and difficulty in getting credit.

6. What do auto industry executives believe about the price elasticity of the demand for automobiles?

7. The executives suggest policies for the government. How would these policies operate to solve their problems, in terms of your supply/demand diagram?

8. How does the present strategy of the industry—cutting costs and production—solve their problem in terms of your supply/demand diagram?

I-11 C **Cash Rebates** Toronto *Globe and Mail*, January 8, 1976

CHRYSLER CORP. PLANS REBATES UP TO $400 TO CUSTOMERS IN U.S.

DETROIT—Chrysler Corp. is introducing cash discounts of up to $400z in the first major price reduction in the auto industry during the current sag in car sales.

The firm told a meeting of car dealers yesterday that it will kick off an unprecedented $5-million promotional campaign on television Sunday during the Super Bowl football game. The campaign, called a Car Clearance Carnival, will feature discounts on a different car each week. It also included light trucks and vans.

(A spokeman for Chrysler Canada Ltd. of Windsor said the company has no information on such a program for Canada and it is strictly a U.S. affair.)

The Ford Motor Co. is also expected to react with a plan soon to counter sagging sales. New car sales are down more than 25 per cent industry-wide since record price increases averaging about $450 a car took effect on new models in September.

The five-week Chrysler campaign, scheduled to run from Jan. 12 to Feb. 16, offers rebates of $200 to $300 on a specific model each week. Another $100 will be rebated on specific trade-in models that dealers said will include both Chrysler and competitors' cars.

An announcement of a Ford pricing plan for its cars was expected today. Sources said Ford probably will offer discounts on optional equipment. Like Chrysler's plan, this would not roll back regular sticker prices.

A General Motors spokesman said GM is not expected to announce a new pricing program this week, although a sales incentive plan in effect at Chevrolet allows dealers to discount prices.

General Motors is also reported to be considering bringing out a stripped-down lower-priced version of its Nova compact.

Dealers said the Car Clearance Carnival marks the first time in automotive history that rebates will be given to buyers directly by the company and not by the dealers.

The campaign allows the company to concentrate on those cars of which it has the biggest backlog. It also allows it to zero in on customers with older cars from competitors and make it attractive for them to switch to a Chrysler product.

Chrysler, which last month closed five of its six assembly plants and laid off half its white collar workers in an effort to reduce its huge backlog of cars and reduce costs, has, in the past two months, seen its share of the total car market drop from 16.7 per cent to 13.2 per cent.

---

1. Suppose the average price of a car before the price increase was $4500. Use the information in this article to calculate the approximate price elasticity of the demand for automobiles. Since the $4500 figure is probably too low, in what direction would a higher average price for cars push your estimate of the price elasticity?

2. How does your estimate from (1) above compare with the executives' opinion (from question (6) following the second article)? How would the executives respond to this comparison?

3. Why do you think the program is not being extended to Canada?

4. Is it rational for the auto producers to offer rebates now even though they may be certain that sales will eventually rebound? Explain.

5. Why would it be Chrysler, rather than GM or Ford, who took the lead in cutting prices?

I-12 Regulated Monopolies

Textbook theory of monopoly behavior shows that too little is produced at too high a price, with excessive profits, relative to the quantity, price, and profits associated with a perfectly competitive industry. Because of this result, governments often regulate monopolies to ensure that they do not exploit their monopoly position. Such regulation usually takes the form of quantity regulation (forcing railroads to run certain routes, for example), price regulation (controlling hydro prices, for example), or both.

Price regulation is tricky, because it is difficult to find the "correct" price. There are two reasons for this. First, it is difficult to get the firms in question to provide the proper cost figures, leading to endless quarrels over what the proper figures really are, as witnessed in one of the articles below. Second, it is difficult to convince governments of the relative economic merit to society of different pricing rules. For example, the government could force the firm to: (a) price so as to have marginal cost equal marginal revenue; (b) price at minimum average cost; (c) set price equal to average cost; or (d) set price equal to marginal cost. Which of these alternatives would you choose were you the government?

Rule (a) is what the monopoly would do, if it were unregulated, to maximize profits. The theory of monopoly has already told us that this creates too high a price and not enough output. Rule (b), although it ensures production at least cost, disregards whether or not that output is wanted. Why produce 5 million frisbees of which 4 million are unwanted, just because 5 million happens to be the output at which average cost is least? Rule (c) is the rule usually adopted by government regulating agencies—it provides the good or service at a break-even price (here "break-even" includes a reasonable return on profit to the firm). Rule (d) is the rule usually advocated by economists. Although this rule does not guarantee a "break-even" operation (implying subsidies or extra profits taxes by the government) it does create the "right" quantity; rule (c) does not guarantee the "right" quantity.

What is the "right" quantity? The right quantity is that quantity for which the cost of producing the last unit equals the benefit provided by that last unit. If marginal cost is below marginal benefit, society's welfare can be increased by producing more of the good or service in question; if marginal cost exceeds marginal benefit, society will gain by reducing output. Since the price of a good measures its benefit (i.e., if the benefit were lower than the price, consumers would spend on other goods, reducing the demand, and thus ultimately the price of the good in question), price should be set equal to marginal cost.

If this creates an excess profit for the firm, the government may simply tax it away with a lump-sum tax. If it creates a loss for the firm, however, the government is faced with the less acceptable alternative of providing a lump-sum subsidy. Why should providing a subsidy be better than making the firm price at a break-even level? If the firm priced at the break-even level there would be a higher price and a lower quantity; marginal cost would be less than marginal benefit—too few of society's resources would be allocated to this industry. If price is set at marginal cost, and a subsidy is provided, resources will be taken from a wide variety of industries in which marginal cost equals marginal benefit, and applied to this industry, in which marginal cost is less than marginal benefit. The result is a net gain to society.

All of this assumes an existing stock of capital (such as railroads and locomotives) for which the fixed costs would be incurred in any event. If production of the "right" output involves an increase in the capital stock, consideration of total rather than marginal costs must be made, since adding capital stock is now in itself a marginal change (on a larger scale).

I-12 **A Railroads** Vancouver *Province*, April 28, 1976

Former Liberal minister lambasted by Tory MPs

Special to The Province

OTTAWA — Two Alberta MPs Tuesday launched a vigorous attack against Canadian Transport Commission (CTC) president and former Liberal cabinet minister Edgar Benson.

Both Jack Horner (PC-Crowfoot) and Don Mazankowski (PC-Vegreville) tackled Benson on controversial railroad aspects when he appeared before a standing parliamentary committee.

Mazankowski, continuing a long time Western fight against railroad branchline abandonment, said government inquiries were using "misleading and distorting" costing figures to show that certain branchlines were economically unviable and should be shut down.

The Vegreville MP charged that on-line and off-line operating costs were being grouped together despite the fact that if the branchline was abandoned the off-line costs, those relating to mainline grain shipment operations, would still remain even though farmers had to haul their grain to the mainline by some other means.

Surely, insisted Mazankowski, if branch lines were ging to be saved a fair assessment of their economic viability should be done?

Benson said the CTC had made all kinds of information available to commissions investigating branchline abandonment and grain hauling.

But, while not actually disputing Mazankowski's argument, he said Ottawa had given the various independent commissions the job of evaluating the two problems and not the CTC.

Horner, front bench opposition transport critic, charged that some branchlines were in such poor condition that trains could drive over them at no more than 10 miles per hour.

Meanwhile, in Montreal, a bus company told the CTC that the subsidization of rail passenger service inhibits the development and expansion of a passenger bus industry through the creation of artificially low fares.

The railways are charging artificially low rates for their passenger services because the federal government subsidizes 80 per cent of their deficits on those services, Provincial Transport Enterprises Ltd. said in a brief to a hearing investigating the trans-continental railway system.

Back in Ottawa, a Toronto Liberal MP charged that Transport Minister Otto Lang has raised a "red herring" in statements that it would cost $1.5 billion to improve tracks to carry passenger trains at speeds of more than 100 miles an hour in the heavily-populated corridor between Quebec City and Windsor, Ont.

1. Reinterpret Mazankowski's "on-line and off-line" argument in terms of economists' jargon of fixed costs, variable costs, average costs, and marginal costs.

2. What should a fair assessment of the economic viability of a branch line involve? Relate this to basic principles concerning profit maximizing by firms.

3. Of what relevance to the arguments reported in the article is the fact that trains could drive over some branch lines at no more than 10 mph?

4. The article notes that a bus company is complaining about the subsidization of rail passenger service. What effect does this subsidization have on the bus company's demand or supply curves?

5. Some people claim that buses use the public roads for commercial purposes free of charge, whereas the railroads must pay to maintain their tracks, and that this constitutes a subsidy to the bus companies. Comment on the relevance of this argument.

6. Consider the railway as a monopolist on a branch line, represented graphically in Fig. I-12.1.

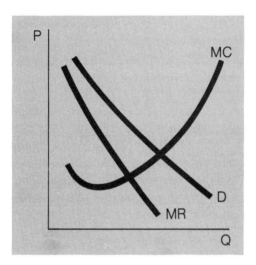

Figure I-12.1

a) What price will the railway charge, with what resulting quantity of service provided?

b) Draw in a representative average cost curve to reflect Benson's position.

c) How can you represent on the graph the profit or loss associated with operating at the position determined in part (a)?

d) How does a government subsidy of 80 percent of the railway's loss affect its price? Does this agree with the opinion of Provincial Transport Enterprises Ltd.? If not, how would this company defend itself?

Danger! Low voltage may unplug Ontario

By Alan Pearson

THE ENTIRE rate structure for the pricing of electrical energy in Ontario may radically change if the directions indicated by the preliminary findings of **Ontario Hydro's** costing and pricing studies are confirmed and implemented.

Some of the changes would alter the lifestyles of some categories of workers, while others could bring startling changes in costs to users.

"There is no relief in sight on rates, and a 22% increase (now in place) is at the low end of what (Ontarians) can expect over the next few years if the lights are to stay on," says Robert Taylor, chairman of Ontario Hydro.

If Hydro doesn't get the rates it wants, Taylor says, "the deterioration in service and reliability will insidiously commence and inexorably continue."

This rumble of distant thunder rolled across the energy field last week when Taylor and Gordon Davidson, director of Hydro's costing and pricing studies, spoke to the Ontario Municipal Electrical Association and the Association of Municipal Electrical Utilities, in Toronto.

There is a threat of electrical shortages in Ontario within five years if current growth in demand is maintained, Taylor warned. He stressed the necessity for stringent conservation measures. He pointed out that a month ago capital funds were further curtailed, and this, he said, means deferment of power stations, transmission facilities, and nuclear-power developments.

Taylor said it has become more difficult to forecast demand for electrical power because of the nature of current conditions and attitudes.

Further, he said, higher costs of primary energy and nuclear-fuel financing, construction and changed cost patterns in power production make current pricing policies progressively more unworkable, even dangerous.

Taylor had other doleful tidings: if voluntary conservation fails and consumption continues to **grow annually at the accustomed 7% over the next decade, blackouts might be expected.**

By 1985, he said, the people of Ontario must have reduced their expected growth in consumption "by two Niagaras."

Davidson argued the case for a fundamental change in the rate structure. He pointed out that **the findings of the costing and pricing studies, not yet completed, do not necessarily represent the policy thoughts of Hydro's board and senior management.**

Davidson said one of the major innovations in Hydro's early development was the rate-structure design, which was promotional in nature. The value of economies of scale were discovered and this later led to quantity discounts for electricity. Hydro soon discovered that the cheaper the cost of electricity, the more demand grew. In the early decades of the century, this helped lead to the economic expansion of the province. But today, he said, the sustained high growth of Ontario Hydro is giving rise to alarm, not only to environmentalists, but to Hydro management.

Management is concerned, for example, about the higher costs of primary energy and construction and the latest development, the ceiling placed by the Ontario government on Hydro's borrowings, despite the persistent need for new generating plants, Davidson said.

The Hydro costing and pricing studies examined some of the impacts that are likely should the new rate structure be implemented — the impact on the consumer price index, location effect on industry, the effects on employment, usage patterns of electricity, and the effect on various social and industrial groups. The conclusion reached:

"All in all, major dislocations do not appear likely for the remainder of this decade, since in most cases electrical costs will remain, even after price increases, a relatively small component of the value added in industry and commerce."

Further, because more women are entering the work force, their income will serve to "dampen considerably the impact of increasing electricity prices."

The objective of the new system of rates would be to try to redistribute the periods of heaviest demand for electricity, both during the day and the year. Hydro would be delighted, for instance, if half the electricity users were to use it at night rather than during the day. Another objective would be to accommodate the demands of conservation critics.

Conservationists are concerned primarily about energy waste, Davidson said.

As far as Hydro is concerned, he said, waste is "any price below what it costs you to produce **another unit of electricity, or what you save if you produce one less unit of electricity . . . That is, any price below marginal cost . . . leads to wasteful use of electricity."**

Marginal costs are "those costs which are about **to be committed by Ontario Hydro on behalf of our customers to the production of electricity."**

In view of this, the Hydro studies suggest that the utility change from traditional average-based rates to rates based on expected costs of production for additional usage. The result of this would be lower **bills for the prudent user, higher ones for the lavish users in all sectors.**

One of the aims of the study is to try to get users off peak-period times, when power is more expensive to produce. The preliminary findings also suggest time-of-day metering for the large user.

This part of the study says that most users can come out with lower power costs by careful planning of power use.

Time-of-day pricing is not a new idea; both New York State and the State of Vermont have it under active review and possible implementation. Also, British and French electrical utilities, which have had it for 20 years, have each estimated annual savings of $250 million by instituting marginal time-of-day pricing.

The Hydro studies team is conducting a cost-benefit analysis of the feasibility of abandoning bulk metering such as is now used in multi-family dwelling units, since such a move would yield significant conservation benefits. It was discovered that residents in bulk-metered apartments use on average 34.5% more electricity than when individually metered.

Davidson argued that implementation of the new rate system is the only way — short of government fiat — that at present can be seen to contain growing demand for electricity.

The new rate structure, if adopted, Davidson said, would give the customer the opportunity to make his own cost-minimizing calculation and act accordingly.

The industrial and commercial user will feel the biggest impact if the new rate structures ever go into effect. They will have to consider shifting their electrical demands to off-peak periods. Single-shift companies may have to consider double shifts, and three shift operations may have to reschedule so their electrically intensive equipment is used during the night.

The battle for higher rates is obviously on, at least as far as Ontario Hydro is concerned. Next, consumers should hear the government's case for slower growth, hopefully buttressed with fresh forecasts of likely consumption.

---

1. The second paragraph means little to the reader until the entire article has been read. Explain briefly the meaning of this paragraph.

2. What is meant by the statement that "higher costs of primary energy and nuclear-fuel financing, construction and changed cost patterns in power production make current pricing policies progressively more unworkable, even dangerous"? How could they be dangerous?

3. Explain the role played by economies of scale in the development of Ontario Hydro. What is the current status of economies of scale?

4. What arguments are presented in the article to suggest that the price elasticity of the demand for electricity is low?

*5. Using the contents of the article, calculate an estimate of the price elasticity of the demand for electricity.

6. Why would Hydro be delighted if half the electricity users were to use it at night rather than during the day? How do they suggest this be accomplished?

7. The article states that power is more expensive to produce in peak periods. What does this imply about the nature of the firm's cost curves at the current output level?

8. What evidence, if any, is presented to defend the statement that "any price below marginal cost . . . leads to wasteful use of electricity"?

9. Explain the nature of "rates based on expected costs of production for additional usage." How do they differ from the present rate structure? How would they affect the normal household?

10. What is a cost-benefit analysis? Explain what would be taken into account in undertaking such an analysis for the abandonment of bulk metering.

11. Ontario Hydro is a regulated monopoly, and as such must have its rates approved by government. Suppose the monopoly can be represented by Fig. I-12.2, and assume that demand must be met (i.e., no blackouts permitted).

    a) Suppose the firm were unregulated. Use the diagram to indicate the price level, output level, and profits of the firm.

    b) Suppose the government wished to set price so as to maximize the quantity provided by the firm. What price, quantity, and profits would result?

* For more advanced students.

c) Suppose the government wished to maximize the output of the firm, subject only to the constraint that the firm not suffer a loss. What quantity should it require the firm to produce? What price and profit would result?

d) Suppose the government decided to set price so that the marginal utility of the last unit of electricity consumed were equal to the marginal cost of producing that last unit. What price, quantity, and profit (or loss) would result?

e) Suppose the government decided to set price so that rates were "based on expected costs of production for additional usage." How would you structure such rates in terms of the diagram? What effect would they have on total demand and profits?

f) From the contents of the article, which price-setting rule is presently being employed? Which is being recommended by Ontario Hydro? Which do you recommend? Would you supplement your rule with any further regulations, subsidies, or taxes? Explain.

12. Many regulated industries are "natural monopolies" and would be represented by Fig. I-12.3 rather than by Fig. I-12.2.

a) Describe in words the main difference between these two monopolies.

b) Explain how the "natural monopoly" gets its name.

c) Answer parts (a) through (d) of question (11) above using Fig. I-12.3 in place of Fig. I-12.2.

d) What is the most significant difference between the ordinary and the natural monopoly that you encountered in answering part 12(c) above? What policy implication does this difference have?

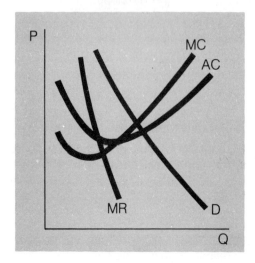

Figure I-12.2

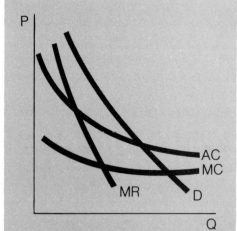

Figure I-12.3

I-13 Taxing a Monopoly*

As noted in the preceding section, governments, in their role as protectors of consumers, often regulate monopolies to prevent them from taking advantage of their captive customers. This is particularly prevalent when the government itself is responsible for creating the monopoly situation in the first place, as is the case in the example described in the article below.

Government regulation can take many different forms. Government can set the price and allow the firm to determine the quantity, it can set the quantity and allow the firm to set the price, or it can control both price and quantity, utilizing taxes or subsidies as supportive measures. These kinds of regulatory activities, however, usually must be specifically sanctioned by law. In general, the government has at its disposal only the less direct regulatory mechanisms of taxes or subsidies; the government may apply a tax or subsidy of some kind, but the firm is free to set price and quantity as it chooses.

Different types of taxes have different impacts on price and quantity, as the questions in this section illustrate. A common tax in this context is a "rent." "Economic rent" is the return to a factor over and above the return necessary to prevent the factor from transferring to some other use. (Monopoly profits are thus often called economic rents.) It can therefore be taxed away without distorting the allocation of factors in the economy. Most taxes called "rents" are designed to tax away an "economic rent."

* Suggested for more advanced students

I-13 A Freeway Gas Stations

Toronto *Globe and Mail*, May 14, 1976

Lower rents proposed to cut gasoline prices at pumps on freeways

By PETER MOSHER

The Ontario Government may lower rental charges to oil companies operating service centres on freeways if they guarantee to pass the savings on to consumers.

Transportation Minister James Snow told the Legislature yesterday one reason gasoline prices are higher than elsewhere at 23 service centres on Highways 400 and 401 is the leasing arrangement the companies have with his ministry.

The companies' 25-year leases are based on a percentage of their gross revenues. That means they pay more as gasoline prices rise, and the Government is now taking in what Mr. Snow later called "a windfall profit out of motorists who are already hit" by high gasoline prices.

The percentages paid the Government are as high as 20 per cent Mr. Snow said, although 10 per cent was a more usual figure. The companies bid for the right to set up stations on the highways, and they have overhead expenses other stations don't have, such as the requirement they be open every day for 24 hours and provide emergency vehicles.

Outside the House, Mr. Snow said ministry officials were talking with oil company officials about an alternative. But he said he had no intention of lowering lease rentals "unless there's a guarantee the saving will be passed on to the motorist."

Microeconomics 57

Suppose Fig. I-13.1 represents the demand and cost situation for a given location, exclusive of the rent paid to the government. Note that because of the location advantage, the service station can be considered a monopoly.

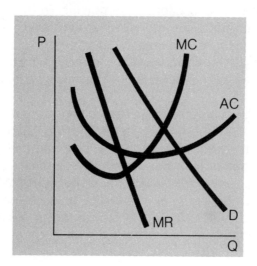

Figure I-13.1

1. Rent is usually paid as a lump sum, independent of the amount of revenue collected by the enterprise. Explain how this would be represented in the diagram and note the influence it has on price, quantity, and profits.

2. Rent could be prescribed as a percentage of profits. Explain how this would be represented on the diagram and note the influence it has on price, quantity, and profits.

3. The Ontario government has chosen to charge rent as a percentage of revenue. Note that this is equivalent to a sales tax; it can be analysed in terms of its influence on price alone. Explain how the effects of this kind of rent can be represented on the diagram and note the influence it has on price, quantity, and profit.

4. Using your answer to question (3) above, explain in words how a potential service station operator would use this diagram to calculate the highest bid he would make to obtain the rights to the station.

5. The Ontario government wants a guarantee that a rent reduction would be passed along to the motorist. Would it be passed along if the government did not force them to pass it along? Explain why or why not.

I-14 Price Discrimination

The term "price discrimination" refers to selling the same item to different customers at different prices. Children's rates for movies and excursion rates for airline flights are examples. Price discrimination is practised by firms because by discriminating they can raise their profits. By examining this phenomenon more carefully we should be able to explain more adequately firms' pricing behavior. Although the word "discrimination" is a value-loaded term, it is not necessary that the practice of price discrimination be a bad thing. For example, although some people end up paying more for the good or service, other people, often those least able to afford it, end up paying less. It is also possible that the extra sales and revenue made possible by price discrimination could turn an unprofitable business into a profitable one, allowing society to enjoy its output rather than go without. As a last example, consider the case for charging different prices for off-peak versus peak-hour electricity, discussed in section I-6. Here there were distinct cost savings to society resulting from the discriminatory pricing.

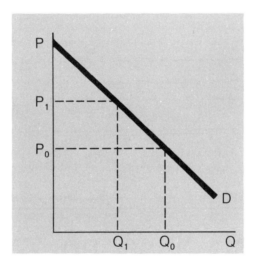

Figure I-14.1

Selling a good at a single price to all customers is a good bargain for those willing to pay more for that good. In fact, anyone represented by that portion of the demand curve above the prevailing price falls into this category. In Figure I-14.1, if the price of airline tickets were P_1 then Q_1 people would fly; if the price were P_0, a larger number of people, Q_0, would fly. Note, though, that the Q_1 people who were willing to pay P_1 form a part of the Q_0 people. These Q_1 people now pay the airline only $Q_1 \times P_0$ dollars whereas before they paid $Q_1 \times P_1$ dollars, a difference shown by the upper left-hand rectangle in Figure I-14.1. They receive what is called "consumers' surplus"—the difference between what they would have been willing to pay for the service and what they actually paid. (Calculating this consumers' surplus on a person-by-person basis should allow you to show that the total consumers' surplus created by price P_0 is given by the triangle above P_0 and to the left of the demand curve.)

Being profit maximizers, businessmen are always trying to find some way of transferring some or all of this consumers' surplus to themselves. If the businessman is a monopolist, and can prevent resale of his output, price discrimination is a method of accomplishing this. Perfect price discrimination exists when the firm can extract from each customer the highest price he or she is willing to pay. For practical reasons this is seldom accomplished. Most price discrimination takes the form of classifying potential customers into groups (as mutually exclusive as possible) having different demand characteristics, and charging each group a different price. Children's rates for entertainment events or for transportation are examples of this.

The airlines are traditional price discriminators, using either the "excursion" versus "regular" fare or the "first-class" versus "tourist-class" fare. The two articles in this section relate to this problem of air fare discrimination. Note the important role played in both these articles by the concept of opportunity cost (the net revenue foregone by not employing a resource in its most profitable alternative use).

I-14 A First-Class Travel

Toronto *Globe and Mail*, August 13, 1976

First class no more

First-class service simply doesn't pay, so Air Canada, finally, has decided to virtually eliminate first-class seats on its short-haul domestic flights. A refreshingly economical decision.

According to Air Canada president Claude Taylor, the airline will remove the first-class seats from its 52 short-to-medium range Douglas DC-9 aircraft and will reduce the number of first-class seats on its Boeing 747 jumbo jets and Lockheed Tristars. The passenger capacity of each aircraft will be increased by replacing the discarded first-class seats with economy-class.

It is no secret—and officers of Air Canada have themselves admitted—that first-class seats don't pay their own way. They have to be subsidized through economy-class fares. Mr. Taylor explains that the purpose of the airline's program of reducing the number of first-class seats is to increase the productivity of each of its planes as much as possible—these are hard times for airlines. If that is so, why keep first-class at all, on any route? It certainly isn't a necessary service.

If Air Canada is really serious about halting its luxurious descent into the red (it lost $12-million on its operations last year), then it should either raise first-class fares to a level that truly reflects their cost or scrap them completely. Few passengers could afford the former. Which leaves the latter.

1. The difference in cost between a first-class ticket and a tourist-class ticket is considerable, far more than the cost of the extra liquor, food and service that goes with first class. How could Air Canada deduce that "First-class service simply doesn't pay"?

2. The article asks, "If that is so, why keep first-class at all, on any route?" Answer this question.

3. Comment on the article's suggestion that "it should either raise first-class fares to a level that truly reflects their cost or scrap them completely."

I-14 **B Excursion Rates**   Toronto Star, March 6, 1976

Money-losing Air Canada finds that if the price is right even an all-night flight to Florida can really take off

GATE 91 is at the far end of lengthy Terminal Two at Toronto's International Airport and it takes a good, stiff walk to get there. But so many people are making the long hike these days that Gate 91 has become one of the brighter spots in the generally dreary state of the North American airline industry.

That's because each Friday and Saturday nights, these days, several hundred economy-minded travellers crowd the departure lounge at that gate about 10 p.m., waiting to board Air Canada's new Nighthawk service to Florida.

The Nighthawk flight is getting a lot of attention from airline insiders, who have wrestled for years with how to make a dollar when your aircraft are flying half-empty much of the time.

The service uses some of Air Canada's biggest jet planes. But each Friday and Saturday night, for the past six weeks, every last seat has been filled—and all seats have been booked solidly for the remainder of the experimental program up to the end of April. And there are more than enough hopeful standby passengers to fill any booked seat that might become vacant.

Behind this phenomenon is a merchandising technique that's an old—or older—than the merchants of Baghdad. Faced with dwindling business Air Canada's marketing men went out and cut prices—dramatically.

To fly from Toronto to Miami and back on the "Nighthawk" costs $119. To Tampa and back it's $110. That's not far off half the regular economy fares, which range between about $200 and $220, depending on the time of year and the other complex rules that determine scheduled fares. You do have to stay in Florida at least eight but not more than 60 days.

If the scene at Toronto's gate 91 were repeated more often in the airports across North America and Europe the airline story would not be the red-ink saga it has become.

This is the era in which the sure and skilful technology of aviation outstripped the fumbling science of economics. Just at the time when the aircraft builders began to deliver a new generation of huge and efficient jumbo jets, the market began to crumble. Aviation fuel prices rose fourfold. A recession rolled in.

United Airlines, like many other big U.S. lines, is in the midst of quietly refitting its nearly new fleet of big jets. This means cutting back drastically on the high-price first class area and squeezing more seats into the economy class.

The moves are aimed at increasing productivity. First class sections have been running two-thirds empty in recent times, while the economy seats were two-thirds full.

The refitting also creates more seats without the need to put up big money for new aircraft—which means more hard times for the big plane makers.

Air Canada now is flying a good number of aircraft with all-economy seating. Four big "stretched" DC-8 aircraft are being converted in this way.

Government-owned Air Canada, with about 120 big jets working on its routes, suffered a loss of $9.2 million in 1974. It expects to report that it lost $12.5 million in 1975.

Yves Pratte, the Quebec lawyer who had headed Air Canada for several years, resigned in late 1975 and was replaced as president by Claude Taylor, 50, a veteran of the company.

Air Canada and CP Air, late this week, applied to the Canadian Transport Commission for fare increases ranging from 6.9 per cent to 11.9 per cent on domestic routes.

Despite taking part in the application, Taylor of Air Canada still has a long-standing reputation as an aggressive pusher for special low-fare rates to meet charter competition and open new markets without losing existing business.

A half-fare program for young people, aged 12 to 21, stays in effect for this year, on a standby basis, inside Canada and out of peak travel periods.

He is also pushing for a new advanced booking excursion fare for the Toronto-London trans-Atlantic run as close as possible to the competing summer charter rate of $369. Charter flights at this price level are flying with full loads—which is the way Taylor likes to see his scheduled aircraft take off.

According to airline officials, Taylor wants to keep his scheduled excursion rates within $20 to $40 above the charter rates, depending on high or low season, while other lines want to keep the excursion rate at a premium of $40 to $80 above the charter rate.

A youthful 50, Taylor has built up a hustling young team of marketing specialists, many of them still in their early 30s. One of them is Gary G. Vogan, 33, who has already served two years as president of the Air Canada subsidiary Air Transit which operated the experimental downtown-to-downtown Montreal-Ottawa service.

Vogan came to Toronto last year as regional sales manager and helped on the details of the Nighthawk operation.

The first announcement of the new service was through a press release Jan. 10. This was followed up by one advertisement in major newspapers.

"Within three days we had sold 75 per cent of our seating," Vogan said in an interview. "On the fourth day we were sold out 100 per cent for the whole program to the end of April. So, we cancelled the rest of the advertisements we had planned.

"We knew some of the big charter airlines were planning low-price charters to Florida, which is one of our biggest markets," Vogan said. "Since we are a scheduled line, we felt we should provide regular scheduled service along with low price.

"We couldn't quite meet the low price of some charter aircraft due to differences in capacity—we have a different seating configuration.

"So the answer is the 'night coach'. We use an aircraft that normally sits idle overnight."

Since regular schedules of most commercial airlines are keyed to the needs of their big market — the businessman, who travels mostly morning and evening on weekdays— this pretty well dictated the time slot the Nighthawk would occupy.

It was a departure from Toronto after the Friday afternoon loads of business travellers had been cleared but before the curfew shuts down Malton flying at midnight. The times chosen were 10 p.m. to Miami and 11.15 p.m. To Tampa.

The flying time is "just under three hours," Vogan says. "The aircraft sits for about two hours in Florida then takes on another load to get back into Malton some time after operations start up again when the curfew is lifted at 7 a.m."

This full-load utilization of an aircraft for two full trips, during a period when it would otherwise be standing idle, makes the price cut possible, Vogan said, while still allowing a fair profit for Air Canada.

"It was pretty well an innovation for Canada," he said, "although some U.S. airlines pioneered cut-rate flights such as the New York-Puerto Rico run more than 40 years ago.

"The first question we had on our minds was 'Would the public buy it?'," Vogan recalled. "It was taken up so fast we were almost oversold before we knew it. It was almost embarrassing."

The second question in the mind of the Air Canada marketing experts was about equally important, Vogan said. "This was where are the new passengers coming from? If they are simply your old passengers being carried more cheaply, you are not improving your overall revenues—in fact you may be hurting."

A preliminary survey of travellers using the Nighthawk has been reassuring, Vogan said. Thirty per cent would have used Air Canada, anyway. Thirty-five per cent would have driven. The remaining 35 per cent were made up of those who would not have gone at regular fare or would have gone by another line.

"More surveys are being taken but, if those figures hold true, we can say we are getting 70 per cent new passengers on the new service."

No loud cheers are in order just yet, however. According to the terms on which the Nighthawk was set up, it is an "experimental program," which is scheduled to wind up at the end of April.

The encouraging part is that if it is ruled a success, then the same formula may be used to encourage travel "within Canada," Vogan said. Nothing has been decided, but this could include special rates from Ontario-Quebec, for instance to the East and West coasts.

The same young Air Canada marketing team last year introduced excursion fares, 35 per cent below regular economy rates, at certain times on some long and short-haul routes inside Canada, and a 50 per cent youth standby discount to help fill empty seats in off-peak hours.

Reprinted with permission The Toronto Star.

---

1. From the contents of the article, what can you deduce about the elasticity of Air Canada's demand curve versus the elasticity of the industry demand curve? Does this fit with the textbook explanation of the relationship between these two demand curves? Explain.

2. The article states that the airlines "have wrestled for years with how to make a dollar when your aircraft are flying half-empty much of the time." Comment on the viability of the following solutions: (a) cut prices; (b) cut quantity; and (c) invoke price discrimination.

3. Put into the context of economic theory the probable meaning of the statement that "the sure and skilful technology of aviation outstripped the fumbling science of economics." (Hint: Make use of the corn – hog cycle—see Section 1.4).

4. What examples of application of the price discrimination approach are cited in the article?

5. What role does marginal-cost pricing play in this article?

6. What role does opportunity cost play in this article?

7. If you were running the Nighthawk service, would you set next year's price at the same level, a higher level, or a lower level? Why?

Part II
MACROECONOMICS

II-1 Measuring Unemployment

For years the unemployment rate in Canada has been used as an index of how close to capacity the economy is operating and as a measure of welfare loss. The current character of the Canadian economy, however, is such that the traditional measure of unemployment is thought by many to be inadequate for either purpose. Although there have always been complaints about the way in which the unemployment statistic was computed (there are many reasons why it is an overestimate or an underestimate of what it purports to measure*), the recent complaints relate more to the character of the unemployed rather than the way in which their numbers are measured.

The current unemployed have a much greater percentage of members of the "peripheral work force," consisting mainly of wives and teenagers, rather than skilled adult workers. Using the unemployment rate as an index of capacity utilization (i.e., as an indicator of how much extra output could be produced if the economy were operating with no idle resources) is now misleading because the economy cannot easily move to this capacity without experiencing severe bottlenecks for skilled workers, with consequent inflationary pressure. With this new structure of unemployed, lowering unemployment to traditional levels would imply that the unemployment rate in certain categories (adult males, for example) would be reduced to extraordinarily low levels, too low to permit a reasonable level of frictional unemployment (workers changing jobs or retraining for new jobs), necessary for the labor force to adapt efficiently to progress and changing tastes.

Using the unemployment rate as an index of welfare loss is also misleading in the context of such a large peripheral work force because only a fraction of those currently unemployed represent households without an adequate level of income. Almost everyone at one time or another has moved in and out of the labor force, taking a summer job, working part-time to pay for a big purchase, or quitting school to work for a year. If a larger than usual number of people decide to enter the labor force at roughly the same time, the economy will be unable to adjust quickly enough to accommodate them. Suddenly, instead of these people being counted as students, housewives, or hippies, they will be counted as unemployed. As a result there is great reluctance to treat the current high unemployment rate as seriously as high rates were treated in the past.

Response to this dilemma has been either to look at different numbers for policy purposes (the job creation rate rather than the jobless rate, for example) or to create new measures of unemployment for welfare purposes (counting only those unemployed not lucky enough to have other working family members or sources of income to alleviate the consequences of their own unemployment, for example). The two articles below provide a good perspective on this problem.

---

* For example, many people pick up part-time work while they are looking for full-time employment. They are counted as employed; this makes the unemployment measure an underestimate. As another example, consider those who don't want to work but say they do in order to collect UIC benefits; these "unemployed" make the unemployment measure an overestimate.

II-1 A Dropouts Welcome

Financial Post, February 22, 1975

Dropouts would be welcome...

By Anne Bower

Canada, we are repeatedly told, has by far the fastest-growing labor force in the industrialized world. Time was when this was considered almost an asset.

In the late 1960s, for instance, it was widely felt that Canada's rapidly growing work force held out the promise of economies of scale and an improved capability for financing social programs. Right now, it's considered to be a downright liability — and with good reason.

In any economic slowdown, the unemployment rate rises, first and foremost, because employment growth slows or actually goes into negative territory. But how high the jobless rate goes can also depend very crucially on exactly how fast the labor force decides to grow.

This year's jobless rate won't average as high as it did in either the worst or the "best" years of the great depression (1933's 19.3% and 1937's 9.1%). But a fast growing labor force could easily push this year's jobless rate to an average that tops the figure seen in other postwar recessions (1961's 7.1%, for instance).

Consider a few numbers.

If employment increases by 1.5% this year — about the same as it did in the slow growth years of 1960, 1961 and 1970 — and the labor force grows by 3%, 1975's jobless rate would average 6.8% (vs 5.4% in 1974). If the labor force grows at last year's pace (4.1%), 1975's average jobless rate would come in at 7.8%.

With labor-force growth this year currently being estimated by economists at anywhere from slightly under 3% to more than 4%, it's clear that an average jobless rate of more than 7% is a very real possibility.

Finance Minister John Turner is no doubt hoping and praying that the growth in the labor force will come in at 3% or less, and so should everybody else. Ottawa's task of righting the economy is difficult enough without having to cope with a big increase in unemployment brought about by people piling into the labor force.

If the labor force climbs by 4% or more, as it did last year, the resulting average jobless rate of more than 7% will almost certainly force Ottawa into adopting very expansive fiscal and monetary policies — policies that will fuel inflation in 1976. The rise in the seasonally adjusted jobless rate, to 6.7% in January from an even 6% in December, has, of course, already brought forth a loud call from some for a new budget.

The wide differences in 1975 forecasts of labor-force growth stem from the fact that there is no single view about what will happen to the so-called "participation rate." Labor-force growth is determined by two factors: first, by growth in the working-age population and, second, by changes in the participation rate (the proportion of the working-age population that decides to belong to the labor force).

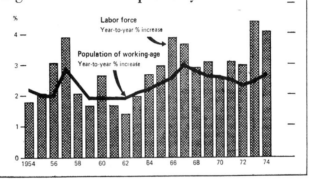

If the growth in the labor force fails to slow from last year's rapid pace, Canada's jobless rate in 1975 will almost certainly reach its highest level in the past 35 years.

The inflow to the labor force attributable to population growth is relatively stable from year to year (see chart). What differences there are from one year to another are largely accounted for by changes in immigration.

By contrast, the inflow to the labor force — or outflow — as a result of changes in the participation rate is not stable at all. The participation rate has been on a long-run uptrend, rising from 52.9% in 1954 to 58.3% in 1974. But year-to-year changes have varied substantially. In 1973, for instance, increased participation accounted for 45% of the 388,000 people added to the labor force, while in 1971, increased participation accounted for less than 20% of the increase.

Labor-force growth has, in the past, tended to reflect the economic situation. In all post-war recession years, except 1960, the participation rate fell slightly, with the result that the growth in the labor force was less than the growth in the base population. This means that in recession years 1954, 1958 and 1961, the unemployment rate was kept from rising even higher than it did by net withdrawals from the labor force. In the mini-recession year of 1970, the labor force grew exactly in line with the base population.

The working-age population is expected to increase by about 2.6% this year. But the labor force could well increase by a much greater amount, as the so-called "additional worker" effect outweighs the "discouraged worker" effect, and sends the participation rate up.

The reason why the participation rate is likely to increase, putting upward pressure on the jobless rate, is that the 1975 slowdown is, in a word, different. What makes it so is that consumer-price inflation is running at 10%-12% compared with past recession years which saw tiny inflation numbers of between 1% and 3%.

To help make ends meet, additional family members (mainly housewives and students) are being drawn into the labor force. Other factors mentioned by analysts as leading to a higher participation rate this year are: the high number of labor negotiations (and, by implication, work stoppages), the fact that women are more committed to the labor force than they once were, and the fact that unemployment insurance benefits make it worthwhile for some to have a short-term job and then remain counted as part of the labor force (and unemployed) when they leave their job and collect benefits.

There will, of course, be some "discouraged" workers as well. Some people will drop out of the labor force because they feel there are no opportunities available and others won't join for the same reason.

Exactly how labor-force growth works out — and what the unemployment rate will be — remains to be seen. But it is clear that the potential is there for a sharper increase in the labor force — and hence unemployment — than has been seen in past periods of economic weakness.

One thing is sure. The labor force is something the Finance minister would be pleased if you wouldn't participate in. Take part in his attempt to reach a consensus on prices and incomes but don't take part in the labor force if you can possibly avoid it.

Macroeconomics 65

1. In which year was immigration to Canada the highest? Explain how you obtained your answer.

2. In which year did the participation rate increase the most? In which year did it decrease the most? Did it remain unchanged in any years? Explain how you obtained your answers.

3. The article states that Turner "is no doubt hoping and praying that the growth in the labor force will come in at 3% or less, and so should everybody else." Why would this be hoped for? Under what circumstances would the opposite be hoped for?

*4. If there is a high level of unemployment, why would the adoption of expansive fiscal and monetary policies fuel inflation in 1976 as the article states?

5. What are the so-called "additional worker" and "discouraged worker" effects?

6. Why would the participation rate, in the past, fall during a recession? Wouldn't the need to make ends meet cause additional family members to join the labor force?

7. Could one legitimately claim that the Women's Liberation Movement has something to do with all this? Why or why not?

8. Is the real problem being faced here one of the level of the participation rate or one of its stability? Explain.

* This question may involve unfamiliar concepts. If so, omit it.

II-1 B Who Are the Unemployed?

Financial Post, August 16, 1975

It's not how many, but who they are

By Sheldon E. Gordon

OTTAWA — The monthly unemployment statistics published by Statistics Canada have long served as effective ammunition for opposition politicians.

A high jobless rate can be pointed to as evidence that the party in power is heartless and callous, as well as incompetent in its management of the economy.

A study to be released late this year by the Economic Council of Canada could help to change all that, though.

Unraveling a paradox

The council, an advisory body with business and union representation, decided to rethink the concept of unemployment in late 1972. It was intent on unraveling the paradox of substantial new job creation coupled with a high and largely unyielding jobless rate.

Part of the exercise involves trying to determine whether the unemployment rate is still a sensitive enough measurement of the degree of hardship inflicted on society by economic conditions.

Council research shows that, with an increasingly large proportion of Canadian families having more than one breadwinner, the unemployment rate may be more legitimate as an index of unused human capacity than of human deprivation.

In an analysis of trends in the job market, the council hopes to reveal subtleties that are not manifest in the monthly labor force survey by StatCan.

Market snapshot

The existing monthly surveys, for example, show nine-10 million Canadians in the job market in any given month.

"But that snapshot doesn't capture the dynamics of the labor market," says Robert Jenness, director of the council study.

In reality, he explains, 12 million-13 million Canadians work during the year, but three million of that total do not show up in the monthly labor force survey because they work only part of the year.

This phenomenon can be viewed another way. The net increase or decrease in unemployment seldom varies by more than 20,000 from one month to the next. But the gross movement of people into or out of jobs can be as high as 500,000 per month, depending on the time of year.

This year, the nation's labor force will grow by about 330,000 people in net terms. But in gross figures, some 2.5 million Canadians — housewives, students, immigrants, and unemployed — will move into jobs. They will be partially offset by flows in the opposite direction.

The "in-out" rhythm of the work force accelerates as the economy becomes more buoyant. Not only do the previously unemployed take the newly created jobs; but women and students who had been, respectively, in the household and at school, also rush into

66 *Dateline Canada*

the job market, and are accompanied by an increased flow of immigrants.

On the margins

Housewives and young people constitute a reservoir of marginally attached workers, says Jenness. An expanding economy sucks them into the labor force, thereby preventing much improvement in the unemployment rate.

But when the economy goes sour and many of these marginally attached workers lose their jobs, they do not immediately pour forth from the labor force to resume their places at home or in school.

Instead, many initially look for other jobs and, failing to find them, swell the ranks of the unemployed. Thus, the marginally attached worker prevents the jobless rate from dipping in good times, and accelerates it in bad times.

These members of the work force are, in council jargon, "supplementary breadwinners." Not they, but their fathers or husbands, bring home the big paycheque in their families. If the marginally attached worker is laid off, the income of his or her family may drop sharply — but no one will starve.

At the peak of the economic recession in 1961, fully 55% of unemployed Canadians were family heads. But in 1973 and 1974, only 35% of the jobless were the main providers in their families. Of Canadian families with someone out of work in 1974, fully 67% had someone else bringing in income — compared to only 55% back in 1961.

Yet despite the changed composition of the jobless ranks, today's high unemployment level is portrayed nonetheless by some as a disaster of Great Depression proportions.

The council study will stop considerably short of trying to develop separate components — one "social," one purely "economic" — of the unemployment rate, but its research could point the way toward such an initiative.

A purely economic indicator would measure unused labor capacity — much the same way that unutilized plant capacity now is quantified. A social indicator of unemployment would then measure the degree of hardship for society caused by various levels of unused labor capacity. Hardship would be expressed in terms of family income.

Underutilization

Thus, the current 7% jobless rate might be redefined as a 7% rate of underutilization of manpower capacity, with resulting material hardship to, say, 3% of the labor force, whose family income tumbles below the poverty level.

Opposition politicians, of course, could still berate the government of the day for economic policies that idle substantial human resources. But that's a lot less dramatic than being able to denounce the rascals for inflicting suffering and misery on a *high* proportion of the working population.

---

1. Explain the difference between the "net" and "gross" movements reported in the article. Is it a good or a bad thing to have a high gross movement? Explain.

*2. The usual specification of the supply of labor has it as a function of the money or real wage. How does the behavior of the article's "reservoir of marginally attached workers" fit into or alter this traditional specification? Explain.

3. What do you think of the suggestion of an unemployment measure with separate components? What policy implications would the adoption of such a measure have?

4. Explain how the concept of the participation rate, as described in the first article in this section, relates to the second article.

* This question may involve unfamiliar concepts. If so, omit it.

II-2 Unemployment Insurance and the Work Ethic

The high unemployment rates of the 1970s stimulated research into the unemployment phenomenon. Two factors that received considerable attention were the work ethic and the influence of unemployment insurance. The work ethic itself received a relatively clean bill of health, as hinted in one of the articles below, and as concluded by a cross-Canada survey. Other studies, however, showed that although people were willing to work, they were not willing to work at unattractive jobs. Much of modern unemployment results, not from people being unable to find a job, but rather from weak job attachment: turnover is very high, more people are quitting than being laid off, and the typical duration of unemployment is quite short. Policies for this dilemma rest on programs to provide on-the-job training and experience for young workers.

The impact of unemployment insurance on unemployment is a different story, however. During the years 1957-1970, average weekly benefits per unemployed amounted to about 29 percent of average weekly earnings, but in 1971 this jumped dramatically to 41 percent. The public began to notice a much larger number of unemployed, particularly among young people, seemingly living off unemployment insurance rather than working. Some called this "cheating," but others noted that it was simply a rational response to changes in economic incentives brought about by the government policies. Newspapers delighted in running articles explaining how the unemployment insurance game "works" (how long you must work before you can collect, how much job hunting effort must be exerted to fool the government official handling your case, etc.) and then running alongside an article quoting employers who are unable to obtain employees, in spite of the high level of unemployment.

Empirical studies of this phenomenon concluded that there is in fact considerable insurance-induced unemployment. In 1972, for example, two percentage points of the Canadian unemployment rate of 6.3 percent were found to consist of insurance-induced unemployed: without unemployment insurance the unemployment rate would only have been 4.3 percent. (Similar results were found for the United States.) This need not be interpreted as a bad thing, however, as the second article below points out.

II-2 A The Work Ethic

Vancouver Province, August 19, 1975

Is 'work ethic' coming back?

Financial Times

TORONTO—Canadians steadily pursued more leisure and less work until 1960, but since then the Canadian work week has fluctuated according to the state of the economy.

A century ago, the average Canadian factory worker slaved 64 hours a week. But by the 1930's, the workload declined to 50 hours, and by 1945 it dropped to 44 hours.

Since 1960, when the average work week in Canadian manufacturing reached 40 hours, it has gone up with economic expansion and declined in recession.

In the boom of the mid-1960s, the work week climbed back to 41 hours. In the 1969-70 recession, it dropped to 39 hours. In the 1972 recovery, it moved back up to 40 hours. In the present recession, the average work week is down to 37.9 hours.

The recent decline may be part of a long-term downward trend which will not be apparent until recovery is in full swing. But even when workers have a shorter work week in Canada or the United States, they demonstrate a marked tendency to choose work over leisure by seeking overtime or moonlighting jobs.

Some resistance is also developing to the other main method of shifting the balance between work and leisure-shortening the work life, the number of years between entering and retiring from the workforce.

When the United Auto Workers union won its "30-and-out" contract in 1973—under which a worker could take early retirement at 55 after 30 years' service—it was considered the wave of the future. But reduction of the work life seems to be following a course similar to the reduction of the work week: after rapid decreases in earlier years, the pace of reduction appears to have slowed.

It was once normal for many Canadians to work from age 15 to age 70, a total of 55 years. By the 1950s, this had dropped to about 45 years.

At 1974 rates of employment, total working life for the average man was 42.46 years. Twenty years earlier, calculations came to 44.1 years.

The same calculations for the average working woman produces completely different results. In 1954, when only 23 per cent of women between 25 and 44 were in paid jobs, the average working life in paid employment was 12.35 years. In 1974, it jumped to 21.12 years.

Nobody knows how much of either the decrease of 1.64 years in the male working life or the increase of 8.77 in the female working life was of choice and how much of necessity.

But the rush of women into the paid work force and the resistance to further reductions of the working life for men owes something at least to a choice of work rather than leisure.

---

1. How could total output and per capita output increase so much prior to 1960 if Canadians pursued more leisure and less work?

2. Why would the average work week go up with economic expansion and decline in recession?

*3. The traditional supply curve of labor, a function of the real wage, is upward-sloping. There exists in theory, however, the case of the "backward-bending" supply curve of labor. Explain how such a backward-bending supply curve could come about.

*4. Does the fall in the work week provide evidence for either the upward-sloping supply of labor curve or the "backward-bending" curve? Explain. What evidence (if any) on this question is provided by the variation in the length of the work week from expansion to recession? The reduction in the number of years worked?

5. Why is the author of the article so careful in the context of women (but not for men) to modify the words "jobs," "employment", and "work force" with the adjective "paid"?

6. How would you explain the "completely different results" for women?

7. One modern economic theory claims that the causal relationship between income and consumption in our modern society runs from consumption to income rather than, as the traditional theory suggests, from income to consumption. How would this theory interpret the last paragraph of the article?

8. Do textbooks define the supply of labor as the number of people wanting jobs or as the total number of hours per year people wish to work? Which is the better way? Why? Which relates most easily to the unemployment measure? Explain.

* This question may involve unfamiliar concepts. If so, omit it.

II-2 B Unemployment Insurance

Vancouver Sun, August 1, 1975

Study defends jobless insurance

By LEONARD CURRY

WASHINGTON (UPI) — Unemployment insurance may increase the incidence and length of joblessness and has certain inflationary attributes, but these drawbacks are outweighed by the positive social factors, according to a new study published by the Brookings Institution.

Brookings Fellow Stephen T. Marston noted in his study data on abuses of unemployment compensation are inadequate and concluded there are social factors in unemployment compensation that far outweigh the liberties a few job seekers take with the system.

Marston said unemployment insurance can cause insured workers to remain unemployed longer than they otherwise might because they can afford to be more selective about the jobs they will accept. Also, it allows a person with minimal job prospects to collect a cheque although technically he should drop out of the labor market until conditions improve.

He mentioned several studies critical of jobless benefits, but concluded there is insufficient information to support charges that benefits actually subsidize unemployment.

There are some subsidy features to the insurance, primarily for jobs that are unattractive or offer seasonal or unstable employment. Without this subsidy, paid by the employer and government, Marston said many unskilled jobs would go begging.

A key factor in jobless insurance, the Brookings Fellow said, is that it prevents employers from taking advantage of workers when jobs are scarce.

Without the guarantee of some income, unemployed workers would be competing for available jobs, which would result in undercutting on pay rates as workers competed for scarce jobs.

Because unemployment insurance helps to insulate employed workers from the effects of job loss, labor costs remain higher than might be expected. As a result, prices remain higher or do not drop quickly.

"In a world of feeble labor demand and limited job vacancies, the inflationary impact of unemployment insurance seems more relevant than its unemployment impact," Marston wrote.

"The system imparts an inflationary bias to the labor market aside from the automatic increase in government expenditures that it causes."

Marston based these conclusions on the economic law of the Phillips curve.

Under this principle, prices fall when unemployment rises. This has not been the experience of 1974-75, a period in which inflation has not abated at a rate sufficient to compensate for the highest unemployment rate of the post Second World War era.

Marston has calculated effects of jobless insurance that suggest the unemployed began looking more seriously for employment when benefits expire.

"After payments are cut off, some workers take jobs they would have rejected or failed to find before that," he said.

"The similar time pattern in withdrawals from the labor force probably indicates that some workers who would have left the labor force held out until their benefits were exhausted.

"This may be only a semantic distinction: The recipients actually may have given up job search before that time, while maintaining a pretense for employment counselors."

On the positive side, Marston said the benefits have prevented wage-cutting wars between job holders and the unemployed competing for scarce unemployment. Benefits also insulate the unemployed from severe economic stress, which occurred in the great depression.

---

1. The author begins his article by admitting three drawbacks of unemployment insurance: (a) it may increase the incidence of joblessness; (b) it may increase the length of joblessness; and (c) it has certain inflationary attributes. Explain why unemployment insurance might have each of these characteristics.

2. In most textbooks unemployment insurance is mentioned only in the context of automatic stabilizers and is thought to be a good thing. What is an automatic stabilizer? How does unemployment insurance act as an automatic stabilizer? Why is it thought to be a good thing?

3. Does the existence of unemployment insurance make fiscal policy stronger or weaker? Explain. (Fiscal policy is a change in government spending or taxing which pushes the economy to a new income level.)

4. The article states that unemployment insurance "allows a person with minimal job prospects to collect a cheque although technically he should drop out of the labor market until conditions improve." What does "dropping out of the labor market" mean? Of what relevance is dropping out of the labor market in the context of this article?

5. Explain how unemployment insurance acts as a "subsidy" to certain kinds of jobs.

6. Marston lists as a benefit the prevention of wage-cutting wars between job holders and the unemployed competing for scarce employment. In what sense is this a benefit? In what sense is it a detriment?

7. Although the article states that there is "insufficient information to support charges that benefits actually subsidize unemployment," other studies do provide convincing evidence to support the charge. What policy implications does this have?

II-3 Consumer Behavior

The consumption function is one of the key elements of the modern Keynesian approach to the explanation of the determination of the level of income. The Keynesian approach explains the income level by explaining the level of aggregate demand for goods and services; consumption is by far the largest ingredient of aggregate demand. The Keynesian multiplier process, explaining how an increase in demand can lead eventually to an even greater increase in income, operates primarily through the consumption function: A stimulus to demand increases income which in turn increases consumption, continuing the expansion.

Keynes introduced the concept of the consumption function in his famous book *The General Theory* in 1936, and defended it on the basis of a fundamental psychological law. Although the simple Keynesian version of the consumption function (in which consumption is a function of income) did fairly well in predicting consumption in the late 1930s, it failed disastrously in predicting post-WWII consumption. During the war, rationing and patriotism led people to generally abstain from consumption and accumulate their savings. At the end of the war, the populace had accumulated considerable wealth and was ready to buy consumption goods in excess of what their current income would have permitted; the Keynesian consumption function should have included the wealth level in addition to income to predict consumption.

Because of this spectacular failure in predicting consumption, economists devoted considerable energy to the development of theories of consumption behavior. The article in this section looks at an important characteristic of a consumption function in an "affluent" society: High incomes lift consumers well beyond the subsistence level and give them "room to maneuver" so that their consumption behavior need not be tied closely to current income.

II-3 A **Consumer Behavior** Vancouver Sun, July 29, 1975

U.S. personal savings highest in 29 years

By JOHN CUNNIFF
AP Business Analyst

NEW YORK (AP) — A rationale for forecasting an upturn in the economy is contained in the soaring personal income and personal savings figures for June, although you'll still find many people with a "show me" attitude.

Spurred by lower income tax withholding and by tax rebates, both figures indicate that consumers now are in a position to go out and buy houses and cars and other big ticket items.

Disposable personal income — after tax income — jumped $63.3 billion in the second quarter, or nearly 10 times the increase that had occurred during the first three months of the year.

Saving soared to an annual rate of $114.6 billion, which meant consumers were able to put away an astounding 10.6 per cent of their disposable income. It was the highest savings rate in 29 years.

A perspective on that level of savings is obtained by comparing it with the rate for other years. During the past 25 years, for instance, the U.S. rate has ranged between 4.9 per cent and 8.1 per cent.

In the view of many economists, including some in Washington, the high level of savings means that the consumer is poised for a buying asault that soon will bring the range back to "normal" levels.

Others point out, however, that the figures also mean the consumer hasn't been buying, although he is in a position to do so. Houses, autos and many retail items remain unsold.

The truth is that nobody can say with certainty just what high level of savings means. Only a study of consumer psychology, of the mental disposition of consumers gives even a hint.

As the dean of consumer psychologists, George Katona, observes, a buying situation is created only when consumers have both the ability and the willingness to spend. It isn't enough to have money; you must have the mood too.

The mood of many Americans is one of insecurity, it seems safe to say. The jobless rate is extremely high. Inflation, while coming under better control, is still a menace in the minds of most people.

To some observers, therefore, the high rate of savings is a measure of insecurity rather than a simple mechanical indicator of ability to buy. People could be saving for a rainy day instead of a buying binge.

In the past, a low level of personal savings often meant that people were taking a chance on the future, that they were optimistic about raises, that they felt financially secure. They were willing to comit themselves.

Confidence in the future is very much an American trait, or it has been. While the U.S. savings rate has been in the single digits since the Second World War, almost no other nation has had that experience.

In Europe, for example, the savings rate in Italy in 1973 was 14 per cent, in Austria 23 per cent, in France 19, Spain 17, Denmark 13. In Japan the rate was around 20 per cent.

The current, extraordinarily high rate for the United States bears close watching over the next month or two. If it remains high it could mean that a high level of insecurity has developed in the country.

---

1. What do you think caused disposable personal income to jump up so much in the second quarter?

2. Most elementary economics courses suggest that an increase in saving leads to a downturn in the economy; the first paragraph of this article suggests, however, that higher saving leads to an upturn in the economy. Explain the apparent contradiction.

3. Can you estimate the marginal propensity to consume out of disposable income during the second quarter? If so, what is your estimate? If not, why not? Can you estimate the average propensity to consume? If so, what is your estimate? If not, why not?

4. Does a large increase in saving imply a change in the marginal propensity to consume or the average propensity to consume? Explain.

5. What policy implications are suggested by the information that the savings rate has ranged widely and that consumers may be thought to embark upon a buying assault or a saving binge whenever the level of savings departs from normal?

6. Economists have developed a consumer attitude index that purports to measure consumers' attitudes towards spending and the economic outlook in general. When the index increases, it is supposed to predict an increase in consumer demand; when it falls, consumer demand is predicted to fall. This article suggests that such predictions will at times fail. Why is that, and how could they be improved?

7. The low savings rate in the United States (relative to some other countries) is interpreted as a result of Americans' confidence in the future. What alternative interpretations could be placed on this empirical fact?

8. Is a high rate of saving a good or a bad thing for an economy to have? Give both sides of the argument.

II-4 The Inventory Cycle

Changes in inventory levels are a key ingredient of investment demand. Most students in introductory economics courses will identify their importance in one of two roles. First, inventories make the "accounting identity" work. In each year total income (what is produced) equals consumption plus investment plus government spending (what is demanded), according to the accounting identity.* But what if the total produced isn't all demanded? Thanks to inventory changes, this cannot happen, since inventory changes are *defined* as being part of business investment. If total demand falls short of the total produced, inventories rise (a rise in investment demand), increasing "total demand," making it equal the total produced.

Second, rises and falls in inventories act as messages to producers, telling them whether they are producing too much (or pricing too high) or producing too little (or pricing too low). Thus the behavior of inventories is a critical ingredient in any explanation of why the economy is doing what it is doing when it is moving from one equilibrium position to another.

In more advanced economic analyses, inventory behavior plays an important role in the dynamics of the economy, particularly with respect to business cycles. This stems not from their size (inventory changes usually comprise less than 5 percent of total investment demand) but from their volatility. The article below explains how inventory changes play this important role.

---

* The accounting identity is usually discussed in textbooks in the chapters on the national income accounts. The two basic ways of calculating GNP, by adding up all income payments and by adding up all sales of final goods and services, creates the accounting identity: total income equals total demand for goods and services.

II-4 A **The Inventory Cycle**

Vancouver Sun, July 8, 1975

U.S. inventories take biggest drop in years

By PETER S. NAGAN

WASHINGTON (CDN) — Many government economists are now raising their forecasts for the second-half business recovery because of the trend shown in some statistics released the other day — new numbers on what's happening to business inventories.

The Commerce Department has just reported that manufacturers' stock of raw materials, goods in process, and finished items fell by 1 per cent in May, the latest month for which figures are available. It was the biggest drop in 17 years.

But it isn't just the sheer size of the May decline that excites the experts. Rather, it is the fact that the figure extends a trend that has been going on for most of the year beyond their previous expectations. The cumulative impact — on top of earlier large work-downs of retailers' holdings — has now reached the point where it has become a more potent force for recovery.

Significantly, the typical business cycle is essentially an inventory cycle. Recessions begin when over-optimistic businessmen overdo their accumulation of goods on shelves in warehouses. They then cut orderings drastically, to work their excess stocks down. The order cuts mean reductions in production and employment — and hard times come.

The cycle reverses when stocks get so low that businesses begin to lose sales because they are out of this particular model, that size, or the other color. That's when they start ordering again, when the factory wheels turn faster again, and when workers are recalled.

History shows that the deeper the inventory cuts, the brisker the rebound in ordering when the time comes. And this is the reason for today's greater optimism.

Many economists had thought that the first quarter of the year would see the deepest inventory run-offs, stocks would still be coming down in the April-June period — but more slowly. Together with the weakness in auto sales, home building, and investment in new plant capacity, this inventory run-down would yield only that sluggish recovery so widely expected.

But now it is clear that the inventory decline continued into the second quarter — and at an accelerating rate. Manufacturers' holdings fell at a $12 billion-a-year rate in April, then $17 billion in May. What's more, the declines have been fairly general — in materials and finished goods; in durables like autos, furniture and appliances; and in soft goods as well.

Because consumer buying has held up surprisingly well during this recession, key measures of the burden of inventories — the so-called inventory-to-sales ratios — have improved markedly.

Over-all, manufacturers' inventories are down to 1.88 times a month's sales, from the 1.96 prevailing earlier this year. For non-durables, the ratio is 2.48, as against an earlier 2.58. And for durables, the figure has dropped from 1.35 to 1.26—the lowest in years, except for the peaks of the late boom.

Economists expect to see some further liquidation of inventories this quarter, but at a slower rate. And that change from rapid to slow — from negative to less negative — shows up as a plus for the economy.

It explains why the analysts expect the recovery to begin this quarter. Many now feel that real Gross National Product — total output of goods and services adjusting out inflation — will be rising at a 4 to 5 per cent-a-year rate this quarter.

But the real pay-off on the sharper than expected inventory run-offs of last spring is expected to come in the fourth quarter. That's when businessmen will feel the spur of bigger gaps on their shelves than they had projected. That's when they will be stepping up their ordering at a faster pace than economists originally thought possible. Hence the raising of the sights. The rate of real growth may now be 6 to 7 percent a year in the October-December period — not the 5 percent or so expected earlier.

The additional employment and overtime that results will, in turn, mean higher incomes and spending. Even the lagging auto and home-building industries could get some further lift.

It may not be such a sluggish recovery after all.

---

1. Changes in inventories usually appear in elementary economics texts at the point at which the reaction of the economy to an increase in spending is explained. What role do they play there?

2. Why would the inventory-to-sales ratio be considered a measure of the "burden" of inventories?

3. Why should a slower rate of liquidation of inventories show up as "a plus for the economy"?

4. Which paragraph in the article suggests that the economy's reaction to the fall-off in inventories will have multiplier effects?

5. Explain the relationship between the business cycle and the inventory cycle (i.e., do their high points coincide, or do they match in some other way?).

II-5 Fiscal Policy

The two major macroeconomic policies used by governments to affect the operation of their economy are monetary and fiscal policy. Monetary policy, discussed in the next section, concerns government manipulation of the money supply in the economy through the Bank of Canada's control over chartered banks. Fiscal policy concerns government manipulation either of tax rates or of its own level of spending. Pure fiscal policy, as distinct from monetary policy, refers to changes in government spending or taxing that do not involve concomitant changes in the supply of money. This type of government policy is usually the first to be introduced to students, via a shifting of an aggregate demand curve, with much being made of the subsequent multiplier effects. The favorite example is an increase in government spending; a tax decrease is another popular example. Both lead to an increase in economic activity through a chain of reactions called the multiplier process. An increase in government spending, for example, increases aggregate demand for goods and services, causing a rise in income. This in turn increases consumption, further increasing demand. This causes another increase in income, perpetuating the process. When this multiplier process has worked itself out, the level of income will have increased by more than the original increase in government spending.

Economists first recognized the potential of fiscal policy after the publication of Keynes's *General Theory* in 1936, but it was not until the early 1960s that they were successful in selling this potential to government leaders. Now, all budget speeches include reference to the impacts of changes in government spending and taxing, and to their multiplier effects. These basically Keynesian ideas, embodied in all introductory economics texts, can also be found in most newspaper articles dealing with macroeconomic policy, as the first article below illustrates.

Monetary policy (discussed in the next section) is also used to stimulate the economy, and has similar multiplier effects. But monetary and fiscal policies differ in several fundamental ways. For example, one policy may be easier to enact than the other, one may affect the economy more quickly or more reliably than the other, and they may have differing impacts on variables such as interest rates.

One major way in which they differ is the extent to which they are discriminatory and the extent to which that discrimination can be controlled. Monetary policy affects the housing industry and small businesses more strongly than other sectors, a discrimination that is very difficult to alter. A potential advantage of fiscal policy, on the other hand, is that its discrimination can be easily controlled, allowing the policy to be used to affect particular sectors of the economy (specific geographical or industrial sectors, for example). In the example discussed in the second article below (the two letters to the editor), fiscal policy had been used to specifically affect manufacturing and processing firms.

II-5 A **Balancing the Budget** *Financial Times of Canada, April 25, 1977*

The pros and cons of balancing a budget

By Susan Goldenberg
Times staff

Ontario's plan to balance its budget by 1980-81 again raises the question of whether governments can or should balance their budgets under current circumstances.

Ontario says it will reduce government spending increases from 11% last year to 9.5% this year, 6.3% in the next two years and 6% by 1980-81. Ontario has had deficits since 1970.

But Treasurer Darcy McKeough was not specific about how the balancing would be done. Treasury officials say spending will be cut by decreasing staff, through attrition, and by reducing or scrapping services.

However, the government has been pursuing this policy for several years, and critics say last week's pledge of a balanced budget is just cosmetics. Government officials deny this. They say their scrutiny of fat in government spending will be even tougher.

With increasing criticism over what is viewed as lack of government spending restraint, the promise of a balanced budget is good politics because the government seems to be controlling its spending.

But economists say balanced budgets can harm an economy which is in the doldrums—the current situation in Canada. Instead, they argue, the economy needs the stimulus of government deficit financing which puts more money into the economy. By contrast, a balanced budget would be more restrictive because it would decrease government spending. Also, to balance a budget, a government might have to raise taxes to bring in additional revenue.

Even if governments slice spending, balanced budgets are not a sure thing because enough taxable revenue may not come in to balance government spending.

For example, 1976-77 Ontario government expenditures of $11.846 billion were $55 million more than expected. Revenues of $10.567 billion were $247 million less. This pushed the deficit to $1.279 billion or $302 million more than anticipated.

"I am not a big fan of balanced budgets," says Carl Beigie, executive director of Montreal-based C. D. Howe Research Institute, a non-profit economics research organization. "This type of mentality got us into the Great Depression. It assumes government does not have the responsibility to stabilize the economy.

"To go to a balanced budget from a deficit budget, according to Keynesian theory, takes the stimulus out of the economy."

---

1. Which action would have the greater impact on the economy, balancing the budget by cutting government spending or by raising taxes? Explain.

2. Why is it that "balanced budgets can harm an economy which is in the doldrums"?

3. Explain what is meant by "the stimulus of government deficit financing which puts more money into the economy." Does it make any difference if the government in question is the federal rather than a provincial government? Explain why or why not.

4. Explain the meaning of the fourth-last paragraph. Is this reaction automatic? Explain.

5. How can government meet its "responsibility to stabilize the economy"?

II-5 B Tax Cuts? *Financial Post*, February 14, 1976

Tax cuts may be needed to boost recovery

By Anne Bower

FEDERAL TAX cuts later this year now seem like a real possibility.

Here's why:

• Right now, trimming down of spending increases and big deficits is the order of the day for all levels of government.

• The danger, though, is not that governments won't rein in enough, but that they will be too restraining as they react to the hue and cry that says big deficits are the culprit behind inflation.

Most provincial governments are enthusiastically slowing down the growth rate in their outlays, and scrambling hard to reduce deficits. Indeed, by cutting back on the increase in payments they make to municipalities, they are, in effect, forcing a mighty increase in municipal tax rates this year and next.

Ottawa, for its part, will probably hold its spending advance to around 14% in 1976, down from around 23% in 1975, and 28% the year before. Meanwhile, despite indexing of the personal income tax, its revenue may be increasing faster than this because of sharply higher unemployment insurance contributions from employees and employers, and because of the progressive nature of personal income tax.

The way things stand now, the turnaround could be too drastic for an economy that is not yet on a comfortably solid path to recovery.

A definite possibility, then, is that Ottawa will have to step in with a tax cut for individuals and companies, to offset part of the new restraint and keep the economy moving up.

Economic Council of Canada Chairman Andre Raynauld is one who thinks so. But, so far, nobody is paying much attention. Raynauld argues that personal income taxes should be reduced — although he's vague, indeed silent, when it comes to details. But the really important part of his advice is this: "Consumption, savings, and investment must be stimulated simultaneously."

If this advice is getting through to Ottawa, and it could well be, it signals a basic shift in the way government goes about its chore of trying to manage the business cycle.

It may really mean, for instance, the last of the big spending by governments. The route that policy makers will probably take from now on will be to do their juggling almost entirely on the revenue side — increasing taxes in boom times, and decreasing them when the economy is operating below potential. Growth in spending will be basically held to growth in GNP.

This would stand in sharp contrast to goings-on in the past quarter of a century.

The years from 1950 to 1975 were marked by a sharply-increasing take by governments of GNP. Such spending rose to 29.7% of GNP in 1960 from 22.1% in 1950, steadied for a few years, and then leaped forward. The share reached around 42%-43% last year, up from 29.9% in 1965.

Ottawa's spending (including transfers to provinces and local governments) came to an estimated 23% of GNP last year vs 15.4% a decade ago (see chart).

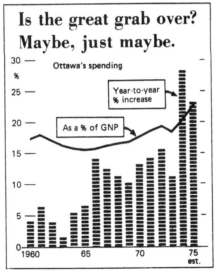

Is the great grab over? Maybe, just maybe.

Just how much of the responsibility for today's big inflation should be placed in the lap of government is a hotly-debated topic among economists. There is, for instance, considerable disagreement about the role of deficits and surpluses in creating and stilling inflation.

But, increasingly, there does seem to be agreement on one point, namely, that the advancing share of governments in the economy has been strongly inflationary. This is mainly because individuals and companies have not been prepared to pay (through taxes) for extra public services or income redistribution and have, instead, demanded bigger and bigger income increases to offset the tax and keep real incomes growing.

McGill University's Christopher Green nicely sums up the policy implications. In an article in the current issue of *Canadian Public Policy*, Green argues that the rate of growth of the public sector should be reduced and so, too, should tax rates.

He puts it this way: "The object would *not* be to attempt to attack inflation by reducing the absolute size of government, or by altering the level of budget deficit or surplus, but rather to combat the cost-increasing tendencies associated with higher taxes and government expenditures *per dollar* of national income."

The inflationary role of deficits, as opposed to growth in spending, is much less clearcut. The Conference Board in Canada's Robert Crozier says, for instance, that there is no evidence to support the view that "deficit financing as such has produced the recent high rate of inflation."

Crozier bravely takes issue with the conventional wisdom, and he makes some telling points. Speaking at a recent seminar sponsored by the Canadian Foundation for Economic Education and Lakehead University, Crozier observed that "we have been running comfortable surpluses in the government account (on a National Income Accounts basis) for the best part of a decade." Chew that one over.

It seems that good arguments can now be forwarded to back up two opposing claims: a budget deficit is inflationary but so, too, is a surplus.

In all this confusion, a balanced budget looks like a good compromise. It could well be that Ottawa will have to come up with a balanced budget in order to restore any credibility to Keynesian fiscal management. It can't — and shouldn't — come up with a balanced budget this year. But it could set its sights on getting there in 1978 or 1979.

---

1. How could big deficits be the culprit behind inflation? Explain by reference to their impacts on an aggregate demand curve.

2. What impact does a tax increase or decrease have on the economy? Explain by reference to their impacts on an aggregate demand curve.

3. Explain the rationale behind the statement that "despite indexing of the personal income tax, its revenue may be increasing faster than this . . . because of the progressive nature of personal income tax."

4. What is the basic Keynesian prescription for managing the business cycle? What modification does the article suggest is in the offing?

5. Does the federal government control the bulk of government spending in Canada? What policy problems might stem from this?

6. Traditional textbook theory says that higher government spending can be inflationary by increasing aggregate demand for goods and services above their supply, i.e., through demand-pull channels. This article suggests that higher government spending also has a cost-push element in affecting inflation. Explain this dimension of government spending.

7. Suppose that Crozier is correct in his statement that the government budget has been in surplus. From your elementary textbook theory, is this consistent with a demand-pull role for government spending? Explain why or why not. (Hint: Review the balanced-budget multiplier.)

8. Evaluate the claim that "a budget deficit is inflationary but so, too, is a surplus," and comment on the article's suggestion that "a balanced budget looks like a good compromise."

II-5 C The Impact of Corporate Tax Cuts

Two letters to the editor, *Vancouver Sun*, June 17, 1975 and July 5, 1975

Report on tax concessions embarrasses Liberals

Sir—On June 5, Finance Minister John Turner finally screwed up his courage to table the long-awaited report on the effectiveness of tax concessions to manufacturing and processing firms for the 1972-74 period. These concessions have since been extended indefinitely and hence require the most careful scrutiny.

It is easy to understand Mr. Turner's delay. The report constitutes one of the most serious condemnations of Liberal economic policy in recent history.

First, it is important to understand how the report was put together. The government last summer sent out questionnaires asking corporate respondents to assess the impact of the tax measures on their firms' sales, employment, investment and prices.

To ask a corporate executive whether he's in favor of tax cuts for his firm, and whether he expects tax concessions to have a beneficial effect, is akin to asking a bear whether he likes honey. The response is largely a subjective one, and the results—one would think—are reasonably predictable.

In view of this built-in bias, it is encouraging—and it is a tribute to the frankness of the businessmen who were polled—that the report can't disguise the fact that the tax concessions represent a waste of billions of dollars.

When the figures in the report are calculated to take into account both those who responded to the survey (67 per cent) and those who did not (33 per cent) but from whom it was subsequently learned that the tax concessions had no impact at all, we find the following: only 32 per cent indicated that investment decisions were favorably influenced by the tax concessions; only 26 per cent indicated that they would be increasing employment because of the tax measures; and only 15 per cent thought that the tax concessions would have any significant effect on stabilizing prices.

The report claims that the tax concessions were responsible for $2.5 billion of new investment. Since the concessions themselves cost the federal treasury $1.8 billion in foregone revenues, the private sector acquired (with only $700 million of their own money) $2.5 billion worth of capital equipment. The enclosed table on capital expenditure raises further important questions both about the government's claim to have stimulated business spending and the state of the manufacturing sector in Canada.

Even after several years of spiralling profits and the injection of $1.8 billion in tax concessions during 1973-74, the manufacturing sector simply managed to approximate the average increase in investment on new machinery and equipment of the four sectors of the economy indicated in the table above.

The government also claims that the tax concession produced a large number of jobs. Again the facts contradict the claim. Consider the table on employment which shows in terms of new job creation, the manufacturing sector, the recipient of the government's tax handouts, has been surpassed by most other sectors in the economy in the last three years.

In fact, we may well ask whether the effect of the public's $1.8 billion was to lessen the manufacturing sector's dependence upon labor—quite the reverse of job creation.

There can be no doubt that the economic stimulus of the $1.8 billion made directly in housing or in the form of tax cuts for low-income individuals, or to increase old-age pensions would have been substantially greater and unquestionably more just.

When the government becomes serious about providing a climate within which Canadian business and industry can thrive, it will, as a minimum, establish firm ground rules for corporate behavior and a tax regime which is both fair and stable; it will not perpetuate wasteful corporate giveaways.

Beyond this, when it becomes serious about creating jobs for Canadians, and meeting other economic and social priorities, it will also have to assume the right to have a major role in all large corporate investment decisions. This is what serious management of the economy is all about.

Arbitrary corporate tax concessions, and superficial reports on their effectiveness are no substitute for a national economic policy.

J. E. BROADBENT, MP
Parliament Buildings,
Ottawa

EMPLOYMENT
Year to Year Per Cent Change By Industry Sector

	1972	1973	1974
All Industries	3.1	5.2	4.3
Goods-producing industries	1.1	5.1	3.6
Agriculture	−5.7	−2.9	1.3
Other primary industries (Forestry, Fishing, Mining)	−3.1	5.1	2.6
Manufacturing	3.5	6.0	2.9
Construction	1.2	9.6	8.9
Service-producing industries	4.3	5.2	4.7
Transportation, Communication, Utilities	4.0	5.9	2.2
Trade	6.0	6.2	5.1
Finance, Insurance, Real Estate	0.0	6.5	8.8
Community, Business, Personal Service	3.6	4.1	4.5
Public Administration	6.3	5.2	5.3

Source: Statistics Canada, Cat. No. 71,001, The Labor Force, January, 1975.

CAPITAL EXPENDITURE ON MACHINERY AND EQUIPMENT
Yearly % increase in spending on machinery and equipment

	1972	1973	1974	1975(ii)
Agriculture and Fishery (i)	23.1	40.3	12.2	15.8
Manufacturing (i)	2.3	23.6	34.2	22.3
Utilities (i)	6.0	33.2	18.6	25.3
Commercial Services (i)	44.7	31.7	14.9	10.3
Average—All Sectors	9.7	27.5	23.7	15.4

(i) These four sectors combined account for nearly 75% of the yearly expenditures on new machinery and equipment.
(ii) Investment Intentions April, 1975 (Statistics Canada, 61-205).

Corporate tax measures have a significant effect

Dear Sir — In a letter published in the June 17th issue of The Vancouver Sun, Mr. Broadbent (NDP, Oshawa-Whitby) commented on the recent report of the corporate tax measures review committee and on the effectiveness of the tax measures for the manufacturing and processing sector.

It is apparent from the report that the economic activity generated by the tax measures has been substantial. In the manufacturing and processing sector over the four years 1972 to 1975, the direct effects were estimated at some $2.5 billion in increased investment, almost 73,000 new jobs, and some $4.8 billion in increased sales, including nearly $1.7 billion in additional export sales.

The impact of the tax measures on the economy as a whole has, of course, been somewhat greater than these direct effects, because of secondary effects on both the sector itself and other sectors of the economy.

Secondary effects on investment, for example, were estimated at about half the direct effect on manufacturing and processing, so that total additional investment in the economy over the four years 1972 to 1975 would be about one and a half times the figures mentioned. The report also estimates that the level of real output in the economy might be in the order of one and a quarter per cent higher in 1975 than it would have been in the absence of the measures, and that the total level of employment by the end of 1975 might be nearly one per cent above what it would have been without the measures.

These are, of course, approximate estimates, but they do show that the tax measures have had a significant impact in terms of generating additional economic activity, and particularly in terms of helping to sustain business investment during a difficult period for the economy.

The primary objective of the tax measures, however, was to achieve a significant longer-term improvement in the viability, competitivenes and growth of Canadian manufacturing. From this point of view, the results are indeed encouraging, in that a significant improvement was reported in the competitive position of Canadian manufacturing and processing firms as a result of the tax measures.

This is a continuing benefit which should provide increased investment, employment and sales well beyond 1975. The report also indicates that the improvement in competitive position is expected to be greater in the period beyond 1975 — as the tax measures have a longer time in which to make their full effect felt — than in the first four years of the measures.

Mr. Broadbent refers to certain figures on over-all investment in machinery and equipment and on over-all employment changes. It may be more meaningful to look at "real" changes (i.e. in constant dollars), rather than the current dollar figures on which Mr. Broadbent's figures are based.

As the table below indicates, real investment by the manufacturing sector on machinery and equipment had declined in the period immediately prior to the introduction of the tax measures, and the level of investment in both 1971 and 1972 was only slightly greater than the level in 1967.

It was this relatively disquietening performance which was the background against which I first proposed the corporate tax measures in May, 1972. In both 1973 and 1974, after the tax measures had become legally effective and had begun to have a significant impact, real investment increased substantially.

Expenditures on machinery and equipment in manufacturing are shown in the following table.

	Constant ($1961)	% change year-to-year
1967	1,577.3	
1971	1,634.2	
1972	1,600.6	—2.1
1973	1,945.0	21.5
1974	2,332.8	19.9

The most recent Survey of Public and Private Investment Intentions indicates that investment in manufacturing is again expected to increase strongly in 1975, perhaps in the order of 20 per cent in current dollar terms. This continuing growth in investment, I might add, is occurring at a time when demand for our exports has declined and when economic conditions have become significantly more difficult.

Growth in the manufacturing sector in the last few years is also indicated by employment data. During the five-year period 1967-71, a total of 927,000 additional jobs were created in Canada, of which 51,000 or 5.5 per cent were in manufacturing During the following three year period, in comparison, over one million jobs were created, of which 229,000 or 21.6 per cent were in manufacturing.

Mr. Broadbent also refers to a figure of $1.8 billion as the "cost" of the measures, and questions whether the benefits obtained were worth the price. The report does in fact estimate the direct effect on federal corporate income tax liabilities as approximately $50 million in 1972, an average of under $600 million in 1973 and 1974, and under $500 million in 1975.

This is, however, only what might be described as the "gross cost" of the measures, before the additional economic activity generated by the measures is accounted for, and before the substantial revenues resulting from this economic activity are taken into account.

The actual net revenue impact, after allowing for offsetting increases in corporation and personal income tax revenues, sales tax revenues and import duties, was estimated in the report as in the order of $30 million in 1972, approximately $400 million in each of 1973 and 1974, and less than $200 million in 1975. The total for the four years was just over $1,000 million. These are necessarily very approximate figures, as the report emphasizes, but they show that the "cost" of the measures in terms of government revenues has been substantially lower than the figures mentioned by Mr. Broadbent.

I would add that these figures for revenue effects do not take into account certain favourable effects on the government's budget position as a result of reduced expenditures, such as reduced unemployment insurance payments, and the net impact on the government's budget balance after allowing for these latter effects would be significantly lower than the figures mentioned.

The corporate tax measures, it should be recalled, were introduced at a time when investment in the manufacturing and processing sector was growing relatively slowly, despite a general cyclical expansion in the Canadian economy. Moreover, incentive programs for the manufacturing sector had been introduced earlier by various countries to which Canada exports and which compete in the Canadian market. The measures were, therefore, intended to stimulate capital formation, to provide new jobs in manufacturing, and to contribute to productivity growth which would help maintain and improve Canada's competitive position.

The report clearly indicates that the tax measures have had and are having a significant effect in terms of improving the longer-term competitive position of Canadian manufacturing and processing (nearly 80 per cent of respondents reported an improved competitive position as a result of the tax measures), and in terms of keeping at home in Canada investment and employment which would otherwise be attracted by tax incentives abroad some 12 per cent of the investment and 17 per cent of the new employment which respondents attributed to the tax measures was expected to have taken place in foreign countries rather than in Canada if it had not been for the tax measures.

JOHN N. TURNER
(Minister of Finance)
Ottawa

1. Evaluate Broadbent's argument in the paragraph beginning "When the figures in the report...". (Turner ignores this argument.)

2. How could the private sector acquire $2.5 billion of capital equipment with only $700 million of their own money, as Broadbent claims? Evaluate Turner's response to this.

3. In which paragraph does Turner refer indirectly to the operation of the multiplier?

4. Evaluate the two authors' arguments relating to the question of "the government's claim to have stimulated business spending and the state of the manufacturing sector in Canada."

5. Evaluate the two authors' arguments relating to the claim that "the tax concession produced a large number of jobs."

6. State in your own words the basic political difference between the economic thinking of these two men as revealed by these letters.

7. When defending political ideologies, very few holds are barred. The most popular technique is to cite statistics favorable to one's own view and ignore others. These letters employ two very different types of statistical procedure. One is to ask the people concerned how they were affected; the other is to measure what in fact happened. Both approaches have weaknesses. What are the weaknesses of each approach? What are their strengths? Both men use each approach when it suits their purpose. Can you illustrate this?

8. From your reading of these two letters, who was the winner of this debate? Defend your answer.

II-6 Monetary Policy

In spite of the fact that Keynes himself stressed the importance and potential of monetary policy, fiscal policy was adopted by the "Keynesian" economists as "the" policy tool, and monetary policy was relegated to a position of minor importance. It wasn't until the 1960s that a revival of interest in monetary policy occurred. This was the result of two things. First, a number of empirical studies appeared, indicating that monetary influences on the economy were considerable and were, in the view of the more controversial studies, more powerful and reliable than fiscal influences. And second, the predictions of the "monetarists" (the defenders of the view that monetary actions are more important than fiscal actions) in the 1960s proved more accurate than those of the "Keynesians."

This revival of interest in monetary policy spurred examination of the way in which monetary policy affected the economy. In the Keynesian view, an increase in the supply of money lowers the interest rate (a rise in the supply of money, relative to its demand, causes the "price" of money, the interest rate, to fall—see Section I-2C) and the lower interest rate stimulates spending (consumers will buy more durables, with a lower cost of credit; business will find more investment projects profitable since the cost of borrowing is less; and municipal and provincial governments will find it financially feasible to build more ice rinks or schools). In the monetarist view, this mechanism is supplemented by general considerations of portfolio equilibrium. An increase in the money supply means that on the whole people in the economy have a wealth portfolio with relatively more money (than other assets) in it than before. To rectify this imbalance, they will draw down their money balances by buying other assets such as consumer durables (directly increasing aggregate demand) or equities (increasing their price and thus lowering the implicit price of capital goods, increasing investment demand).

In the 1970s interest in monetary policy accelerated as inflation accelerated. Ever since recorded time man has noticed that increases in the money supply and increases in prices have gone hand-in-hand. This was spectacularly so in hyperinflations such as in Germany after WWI, in the American Confederate States near the end of the U.S. civil war, and in many South American countries today. And it was also the case in less spectacular instances such as in the seventeenth century when the Spanish discovered gold in the New World and in the second half of the eighteenth century when fractional-reserve banking was invented. True to form, the inflation of the 1970s was accompanied by large increases in the money supply; in Canada the money supply in some years reached annual increases exceeding 20 percent. Although increases in government spending or cost-push forces can create an inflation as easily as can an increase in the money supply, only an increasing money supply can *sustain* inflation. Any inflation involves an ever-increasing demand for money to undertake transactions at higher and higher prices; without an increase in the money supply this higher demand for money will force the "price" of money, the interest rate, higher and higher, eventually causing it to reach heights sufficient to curtail aggregate demand and grind to a halt those forces pushing prices up. It is this consequence of a fixed money supply which explains why governments allow it to grow and thus sustain the inflation: governments do not like to have high interest rates and high unemployment.

As is taught in introductory textbooks, the money supply is controlled by the Bank of Canada, mainly through open-market operations—the buying or selling of government bonds. (The "bank rate," the rate of interest charged by the Bank of Canada on loans to the chartered banks, serves more as a bellwether—a signal—of the central bank's policy.) Purchases of existing government bonds by the Bank of Canada should be sharply distinguished from its purchases of new bonds, however. Existing bonds are held by the public; their purchase by the Bank means that the populace is left holding more money but fewer bonds—the same level of wealth is held, but in a different form. Purchase by the Bank of new government bonds, however, gives the government money which it then spends, increasing the money holdings of the populace with no corresponding decrease in their holdings of bonds—their total wealth holdings are larger. This latter action on the part of the central bank is often called "printing money"; because of the concomitant increase in wealth, this means of increasing the money supply is much more stimulative than is the former means. The inflationary impact of a government deficit is often due to the fact that the deficit is financed by selling bonds to the Bank of Canada rather than by selling bonds to the public.

In the early 1970s the Canadian government was increasing its spending dramatically, often paying for it by having the Bank of Canada print money. This led to large increases in the money supply and substantial inflation. Eventually the Bank of Canada rebelled and embarked on a program of fighting inflation. There were two basic approaches it could have taken. The first option was to sharply reduce the growth rate of the money supply to a rate approximately in line with the sustainable real growth of the economy. This approach was rejected for fear of creating widespread unemployment ("the operation was a success, but the patient died"). This is clear from the words of Gerald Bouey (the governor of the Bank of Canada) in his 1976 annual report: "All sorts of existing arrangements, including virtually all wage contracts, are based on the assumption of some continuing inflation. The attempt to force as rapid a transition to price and cost stability as this prescription involves would be too disruptive in economic and social terms to be sensible or tolerable."

The second option was the one chosen. It is described by Bouey in the same report: "In my opinion it's much wiser to gear down inflation more gradually and over a longer period. Any program to this end that is to have a chance of working must be firmly based on fiscal and monetary restraint, but time must be allowed for the various elements of inflation now built into the economy to be eliminated.... The choice of a rate around which to stabilize the growth of money supply at the present must have regard to the high rate of inflation that now exists in Canada, but the rate of money supply growth should be steadily diminished in the years ahead as a key element in the program to wind down inflation."

This application of monetary policy is illustrative of the use of this policy tool. The two articles in this section look at some of the problems associated with this policy, such as how "money" should be defined, and the need for cooperation between the monetary and fiscal authorities. Later in this book, other articles will further examine this application of monetary policy, recognizing more complicated implications of that policy (ignored in the articles here), such as its influence on the international sector of the economy and its differing impacts on the real versus the nominal rate of interest.

II-6 **A The Money Supply** *Financial Times, July 5, 1976*

Bouey clings to M1 measure —other indicators booming

By Bud Jorgensen
Times staff

OTTAWA — Rates of change in narrow and broad definitions of money supply (see box) have diverged widely in recent months but central bank authorities still are concentrating on the narrow definition (M1) in the anti-inflation fight.

To control inflation, the Bank of Canada has successfully restricted M1 growth to about 10% a year. However, money supply, according to other measures, has been growing at close to 20% a year.

If, as monetarists believe, too rapid growth of money supply is a prime cause of inflation, then increases of about 20% a year seem alarming. But Bank officials say the choice of M1 as a point of control was correct and that the faster growth of money supply according to definitions of M2 and M3 does not signal a return of double-digit inflation.

"We've found it hard to find any specific relationship between the broader aggregates," says the Bank of Canada's governor, Gerald Bouey, about the different rates of change for M1, M2 and M3.

The fact that M2 and M3 have grown so rapidly in the last two years only confirms Mr. Bouey's view that M1 should not be allowed to expand any faster.

Targets for M1

He has set targets for growth of M1. The target between the second quarter last year and the second quarter this year was to keep M1 growth close to 10%.

Mr. Bouey has said he soon will be setting a new and lower target for M1 growth in line with general government plans to trim the inflation rate.

With M1, Mr. Bouey has chosen the most liquid supply of money — cash which can be readily spent or money which can quickly be taken out of the banking system and spent.

The broad definitions of money include less liquid funds — forms of money which are not as easy to spend. Money deposited in banks for a fixed term, for example, carries a penalty if withdrawn early but a higher interest rate compared with a regular savings account if left on deposit for the full term.

Money deposited for a term is not left in bank vaults but is loaned out by the banks (subject to minimum cash reserve requirements). This means that

---

Money supply... and how it grows

% change at annual rates to May

	M1	M2	M3
Last 3 mo.	− 5.7	+ 20.4	+ 19.5
Last 6 mo.	− 7.5	+ 18.2	+ 18.4
Last 12 mo.	+ 10.4	+ 20.0	+ 17.1
Last 24 mo.	+ 8.7	+ 19.5	+ 21.0

Figures seasonally adjusted

M1 — narrowly defined money supply: total Canadian currency in circulation (outside banks, because money is not in circulation if it is on deposit at a bank) plus demand deposits at chartered banks in Canadian dollars. Demand deposits are bank accounts which can be converted quickly to cash and include personal chequing accounts and business current accounts.

M2 — or private money: total currency in circulation and Canadian dollar deposits by the general public at chartered banks. Includes demand deposits, personal savings and all business deposits.

M3 — or total money: M2 plus government of Canada deposits at chartered banks. M3 is about $76 billion. In May, M1 made up 23% of M3 and M2 (which includes M1) made up 96%.

a dollar on deposit can flow back into the banking system one or more times and that dollar can expand money supply.

One of the main lessons to be drawn from the relatively rapid expansion of M2 and M3, Mr. Bouey says, is that people have less confidence in money.

The rapid expansion in M2 and M3 occurred as inflation rates were at historically high levels. Instead of saving, people spent and turned money over quickly in the banking system.

During the inflation of the last three to four years, the money supply (broadly defined) grew most rapidly in 1974. There has been a slight decline in the rate of increase — a rate still high by historical standards — in the last 15 months.

If Mr. Bouey's tight controls on M1 help cut the inflation rate and restore confidence in money, there should be significant drops in rates of growth in M2 and M3.

---

1. The article says that "too large a growth of money supply is a prime cause of inflation." Without referring to numbers, what rate of growth of the money supply do you think is meant by the expression "too large"?

2. The author of the article suggests that 20 percent annual growth of the money supply is too large. In line with your answer to the first question above, what rate would be about right for Canada, to be consistent with a zero rate of inflation?

3. Explain the meaning of Bouey's statement that "We've found it hard to find any specific relationship between the broader aggregates."

4. What is the approximate size of M1 relative to M2?

5. Why would Bouey choose to control M1 for inflation-fighting purposes, rather than M2 or M3?

6. Explain the statement that "a dollar on deposit can flow back into the banking system one or more times and that dollar can expand money supply."

7. How can M1 decrease while M2 and M3 increase? Why might this happen, particularly during an inflation?

8. Explain the statement by Bouey that "One of the main lessons to be drawn from the relatively rapid expansion of M2 and M3 is that people have less confidence in money."

9. Turning money over more quickly in the banking system (third last paragraph) implies an increase in "velocity"—the ratio of GNP to money (M1). One way this ratio can rise is by people spending more to draw down their money balances to a lower desired level. A second way in which this could happen is by people drawing down their money balances by buying other financial assets. Which do you think is occurring here? Defend your answer.

10. As a general principle, why might we expect to notice that velocity rises during an inflation?

11. What is your opinion about using M1 or M2 as an indicator for inflation-fighting purposes?

II-6 B A Rise in the Bank Rate

Toronto Globe and Mail, *March 10, 1976*

Brinkmanship
RONALD ANDERSON

The Bank of Canada's decision to raise the bank rate to 9.5 per cent from 9 per cent, effective on Monday, came as a great surprise to most members of the financial community.

Shortly before the interest rate increase, one investment house had advised its clients to establish a fully invested position in Canadian long-term fixed income securities and in equities because of prospective price gains during the next 12 months. These gains might still materialize over the next year, but in the short run bond prices dropped following the move by the central bank.

Even analysts who had believed a rise in bank rate might be justified by financial developments, notably a rise in bank loans and in the seasonally adjusted money supply in February, were confident that the Bank of Canada would avoid taking an action that would be highly unpopular, in view of the already high level of interest rates in Canada.

Why the raise?

Why, then, did Bank of Canada governor Gerald Bouey choose to raise the bank rate?

It would seem that the reasons were not strictly economic in nature. The economic reasons, in, themselves, are hardly convincing.

In a statement announcing the rate change, Mr. Bouey noted that the money supply (currency and demand deposits) had grown at a rate of 15.5 per cent from the second quarter of 1975 to February, 1976. This is higher than the 10 per cent trend rate of growth of the money supply over the previous two years, Mr. Bouey said.

"Monetary expansion at a rate of 15 per cent a year or higher risks accommodating a rate of inflation in Canada fully as rapid as it has been over the past year."

The growth rate, however, is only marginally higher than the target range of 10 to 15 per cent established earlier by the bank governor.

Not especially good

Furthermore, the base period selected for purposes of comparison was not an especially good choice. The bank did not begin to move toward a restraining policy until the autumn of 1975.

If the money supply growth is calculated from October, 1975, to February, 1976, the seasonally adjusted rate of expansion was about 8 per cent. This, it might be supposed, would be a sufficient degree of restraint, in view of Mr. Bouey's guidelines.

One valid reason for raising the bank rate was the prospect of a refunding issue: A $436-million issue matures on April 1, and the Government is expected to seek some new money, in excess of its refinancing needs.

The central bank probably will have to relax its tight grip on chartered bank cash reserved to accommodate the new issue. A higher bank rate is, in a sense, a trade-off for somewhat less restraint on bank liquidity. It will enable the financing to proceed without the central bank losing control of the monetary situation. If the bank felt a higher level of interest rates was needed to avoid post-financing problems, it could not have waited much longer before announcing the change, without risking market disruptions which might have made debt management more difficult.

Other reasons

But there are other reasons, it would seem, for the bank rate increase. One is that the Bank of Canada is said to be displeased with the chartered banks for being too willing to accommodate loan demand. General loans are estimated to have moved up at a seasonally adjusted rate of 35 per cent in February.

The banks, despite the squeeze on liquidity imposed by the central bank, managed to meet the loan demand by bidding for money in the commercial paper market, by persuading customers to switch from demand deposits to term deposits (which require a smaller reserve base), by running down their holdings of Canada bonds, treasury bills and commercial paper, and by cutting back on call loans.

Some analysts say the Bank of Canada was disturbed by the willingness of the banks to go to such lengths to meet loan demand.

The chartered banks, that is, were not co-operating fully with the Bank of

Canada in its attempt to hold down the expansion of credit.

But, it might be speculated, there is a more important reason for the central bank action on the bank rate. The increase seems to be a message to the federal Government that the Bank of Canada is not satisfied that the mix of fiscal and monetary policies is sufficiently stringent to bring inflation under control.

Significant saving

Mr. Bouey, in common with many other Canadians, appears to feel that a proposed 16 per cent increase in federal expenditures does not constitute an adequate degree of fiscal restraint. The bank governor probably would be much happier if federal expenditures were held down to the nominal growth rate of the economy, which may be no higher than 12 or 13 per cent in 1976. A reduction of three percentage points in a $40-billion budget would represent a significant saving of revenue.

Unless the federal Government adopts a more convincing posture of restraint, the Bank of Canada evidently is prepared to continue squeezing the financial system. The prospect of a steady rise in interest rates and tighter money has the appearance of being a stout lever for persuading the Government to bring down a spring budget that reduce spending and raises taxes, probably personal income taxes.

Higher interest rates would strengthen the dollar and raise unemployment. But Mr. Bouey is determined to slow inflation.

---

1. Why would bond prices drop in response to the Bank of Canada action?

2. The Bank rate is said to be a bellwether. What does this mean? What does a hike in the Bank rate have to do with the rate of money supply increase?

3. What is the meaning of the statement that "the Government is expected to seek some new money, in excess of its refinancing needs"?

4. If the government must refinance $436 million, why doesn't it get the Bank of Canada to wait until after the refinancing before increasing interest rates, so as to minimize its interest costs?

5. What is the central bank's "tight grip on chartered bank cash reserves"? What would relaxing that grip accomplish in the present context? Explain.

6. Why would the Bank of Canada "be displeased with the chartered banks for being too willing to accommodate loan demand"? Isn't making loans the banks' business?

7. What is the difference between a demand deposit and a term deposit?

8. What is the meaning and significance of the smaller reserve base for term deposits?

9. The article states that the chartered banks "were not co-operating fully with the Bank of Canada in its attempt to hold down the expansion of credit." Isn't an expansion of credit equivalent to an increase in the money supply? And doesn't the central bank, the Bank of Canada, have control over the money supply? Clarify this, and comment on the policy problems involved.

10. The article states that the hike in the Bank rate is a message to the federal government to decrease its deficit spending. What power has the Bank to back up this message should the government not decrease its deficit? What would happen if it did not decrease its deficit? (Hint: How does the government finance its deficit?)

*11. How would higher interest rates strengthen the dollar? What impact would this have on aggregate demand and inflation?

12. How would higher interest rates raise unemployment? What effect should they have on inflation?

---

* Suggested for more advanced students.

II-7 Independence of the Monetary Authorities

As should be evident from the preceding section, monetary policy plays an important role in the economy, particularly with respect to the problem of inflation. Because of this, the institutional environment in which monetary policy is conducted is of relevance. One particularly important question in this regard, addressed briefly in the preceding section, is the degree to which the monetary authorities (the central bank) and the fiscal authorities (the government, either Parliament or the Cabinet) should cooperate. This question is often put in starker terms: "Should the central bank be independent of the government?" In the United States, the Federal Reserve Board (the U.S. central bank) is independent of Congress and the President. In Canada, independence was the case until 1967 at which time the ultimate responsibility for monetary policy was placed in the hands of the Cabinet; in practical terms, however, the Bank of Canada (Canada's central bank) still acts independently of Parliament and the Cabinet. There are strong points on both sides of this "Independence of the Monetary Authorities" controversy. The major argument of those opposed to independence is that elected officials should control monetary policy to ensure that the proper mix of monetary and fiscal policy is attained; those in favor argue that elected officials must be watchdogged to guard against financial irresponsibility. This debate among academicians hit the front pages during the early 1960s when James Coyne, the then governor of the Bank of Canada, differed publicly with the government and attempted to assert the Bank's independence. The upshot of the ensuing struggle was that Coyne resigned (in effect he was fired) and was replaced by Louis Rasminsky who was prepared to cooperate with the government in the event of a dispute over Canada's appropriate monetary policy. This article draws on Rasminsky's thoughts to comment on the independence debate and to suggest improvements in the institutional framework within which the monetary authorities operate.

II-7 A The Bank's Responsibilities

Toronto *Globe and Mail*, November 17, 1975

Bank should report regularly to House

BY R. J. ROBERTSON

Mr. Robertson is a member of Storey, Boeckh & Associates, editors of the Bank Credit Analyst.

I HAVE FOLLOWED with great interest and concern the debate generated by J. Douglas Grant and William W. Moriarty (Independent Central Bank Needed to Deal with Dollar, Sept. 22, The Globe and Mail). I use the word concern in light of Louis Rasminsky's disappointing response (Oct. 3) to the serious issues raised by Mr. Grant and Mr. Moriarty, whose key points were:

That the 1967 Amendment to the Bank Act transferred ultimate responsibility for monetary policy from the Bank of Canada to the Minister of Finance and the Cabinet.

Inflation a result

That inflation results when the quantity of money and credit is increased, on a sustained basis, at a rate that is higher than the economy's capacity to produce real goods and services.

That governments increase the money supply at a too-rapid rate because it enables them to finance an expanding government sector without the requirement of increasing taxes or borrowing from the economy's real savings.

That the authors relate excessive expansion of money supply to price inflation and illustrate how the money supply is expanded through central bank actions on chartered bank reserves.

That Graham Towers and James Coyne, the first two governors of the Bank of Canada, were strong-willed and concerned about the bank's independence. However, Mr. Coyne lost out in a dispute with John Diefenbaker, then Prime Minister. Mr. Rasminsky assumed the job with the express and public commitment that in a dispute over policy he would abide by the wishes of the government.

That the authors recommend a more controlled growth rate in money supply, tackling other abuses of the pricing system, educating the public and requiring the central bank to publish minutes of policy meetings and to issue a statement of objectives annually.

Mr. Rasminsky was no doubt hurt by the implication that the governors preceding him were more strong-willed and independent. Perhaps this explains why his reply concentrated on the theme that "there was no basic difference in the philosophy of the first three governors of the bank regarding the ultimate responsibility for monetary policy". Of course there wasn't. They all agreed that in the final analysis the government was, but, of the three, Mr. Rasminsky requested formal recognition of this fact as a condition of employment and during his testimony before the Royal Commission. It is unfortunate that Messrs. Grant and Moriarty used the term "ultimate responsibility"; "formal responsibility" would have been better.

Mr. Rasminsky emphasizes in his letter that directly or indirectly he was never pressured to case monetary policy by the government, but that surely begs the question of whether any pressure had to be exerted in the first place. If one reads the lengthy section of Mr. Rasminsky's testimony on monetary policy before the Royal Commission (pages 101-105) of the Bank of Canada submission the conclusion is inescapable that the governor and the government, largely through the aegis of the finance department, were in extremely close contact. This is illustrated in Mr. Rasminsky's remark on page 101: "There is constantly a process of discussion going on between the bank and the government. When I say the 'government', I refer to the various levels of government. The relations between the bank and the officials of the department of finance are very close; we see a lot of them, and we are constantly discussing the whole range of financial and economic developments, and they know what monetary policy is being pursued by the bank and in the course of those discussions they have an opportunity to express views about the monetary policy.

"In addition to discussions with the officials, I see the Minister of Finance regularly, and the same applies there. The Minister of Finance has the opportunity of expressing any views that he cares to express on monetary policy."

Obviously, in such a relationship the fine lines of where pressure occurs and where a governor can no longer support the government's view and must thereby resign are concepts somewhat removed from the normal workings and settings of monetary policy.

Philosophical debate

Of course, the debate over semantics and philosophy could extend interminably and to little point. What needs to be done is to get the setting of monetary and fiscal policy out into the open so that Canadians can see what is being decided and pass intelligent judgment on it.

As a first step, the central bank should be required to report to parliament rather than to the government. The appropriate committee receiving the written and oral testimony should have the right to question and probe the rationale for the policy setting. The existing system of an annual report of the governor to the Minister of Finance produces little more than a bland statement of generalities so carefully hedged that the intent of monetary policy is totally obfuscated.

Second, while the setting of monetary policy should continue to be the joint function of the bank and the Government, the governor should report projected growth rate targets for the various credit aggregates including bank reserves to parliament on a regular basis such as quarterly. While one should not assume too much knowledge on the part of politicians about the relationship between money supply and inflation one could assume that if an intended growth target of say 20 per cent was announced in advance some parliamentarians might deduce that could produce 5 per cent real economic growth and 15 per cent inflation.

There are difficulties in having the central bank present growth rate targets in advance. Central banks find it much easier to explain events after the fact if they are allowed to operate from a position of secrecy, with little communication with the public. Politically unpopular increases in interest rates might also have to be tolerated at embarrassing times under a policy of published target growth rates.

Obviously, the central bankers could extend the list but these inconveniences to the central bank are well worth bearing to help avoid a continuation of the disastrous monetary-fiscal policy course that Canada has followed in recent years. Strong steps to reverse the headlong slide into a complete debasement of the currency are absolutely essential.

Other concrete steps can be taken as well. A massive educational task is required to acquaint the public with:

First, the size and growth of all types of government spending relative to the growth of the private sector.

Second, the financing of bloated government including the use of the printing press and inflation.

Third, the ultimate loss of all individual freedoms if fiscal-monetary discipline is not restored.

A major effort is necessary, through and by the press, to generate awareness among Canadians of what is happening to their currency as a result of the Trudeau Government's reckless policy course. Retired public servants will be judged as no less faithful if they speak out on all of the issues as well as the philosophy.

---

1. Explain the economic reasoning that lies behind the statement in the third paragraph.

2. What is the meaning of the government "borrowing from the economy's real savings"?

3. How is the money supply expanded "through central bank actions on chartered bank reserves"?

4. What are the implications, in the context of this article, of the difference between "ultimate responsibility" and "formal responsibility"?

5. What does the author mean by "but that surely begs the question of whether any pressure had to be exerted in the first place"? What is he implying?

6. Why might increases in interest rates have to be tolerated under a policy of published target growth rates? What might cause such increases? Explain.

7. To what is the author referring when he mentions "the disastrous monetary-fiscal policy course that Canada has followed in recent years"?

8. What would cause "a complete debasement of the currency"?

9. What does the author mean when he refers to government use of the printing press? (Surely he doesn't mean using a printing press to roll off more dollar bills?)

10. How can inflation be used to finance bloated government?

11. What could the author mean by the ultimate loss of all individual freedoms if fiscal and monetary discipline is not restored? What is your opinion on this?

II-8 Demand-Shift Inflation

In any dynamic, growing economy, the kinds and relative quantities of things that the economy is producing are continually changing. Many would consider this a manifestation of progress. Such a reallocation of resources, however, can create growing pains in addition to the expected benefits of meeting more closely society's wants. A major reallocation of resources in our modern economy has been towards the service sector. Information as to the extent and nature of this growth in the service sector is of use to businessmen contemplating investing in that sector and to students contemplating the nature of their future training. To the economic theorist, however, it is the side effects of such a reallocation of resources that are of primary interest, since such side effects (or growing pains) can help explain some of society's ills (such as inflation) and perhaps suggest fruitful new avenues for policy.

Preceding sections of this book have suggested that the major cause of inflation is excessive money creation. Other sources of inflation exist, however, most serving to supplement the forces related to the money-printing phenomenon. One of these is a combination of demand-pull and cost-push ingredients called "demand-shift" inflation. (Demand-pull inflation is caused by aggregate demand exceeding aggregate supply and pulling up prices; cost-push inflation is caused by increases in production costs pushing up prices.) Although aggregate demand for goods and services may equal aggregate supply, the make-up of the aggregate demand may not match the make-up of the aggregate supply—the things produced may not be exactly what is demanded. As a result, there are some industries experiencing excess demand and other industries experiencing excess supply. According to the demand-shift theory, the excess-demand industries (more numerous when aggregate demand is high) experience inflation which then "spills over" into the excess supply industries. This spilling over occurs through cost-push mechanisms. The excess demand industries experience an increase in their prices, and to attract extra labor they offer higher wages. Those not fortunate enough to be in the excess-demand industries use their market or union power to push up their prices and wages to prevent a loss in real and relative income. This action eliminates the wage and price differential required to induce resources to flow into the excess-demand industries, so the process must begin anew.

Changes such as the development of the oil cartel and the disappearance of the anchovies in the mid-1970s are examples of structural changes in the economic system that would intensify this demand-shift mechanism. This article discusses a less sudden change that has similar implications.

Service-sector growth has made it harder to reduce inflation rate

By Anne Bower

People who work on Canada's auto assembly lines and in textile factories don't need to be told that the 1974-75 economic downturn has been the longest and deepest since the Great Depression.

In many cases, the recession that is now ending has put them or their neighbors either out of work or hovering dangerously close to it.

For a remarkably large number of Canadians, however, it's probable that recession has been something to read about, and little more. Unemployment (or the threat of it) has been almost non-existent for the large segment of the work force that is employed in the service industries.

Implications

The implications of this go far beyond the immediately obvious social problem of hardship for some parts of society existing side-by-side with continuing prosperity for others.

There is considerable evidence that the shift away from goods-producing industries toward the service sector ranks high on the growing list of reasons why the reduction in inflation is a more difficult task for economic policy than in the past.

The service sector is by far the dominant part of the economy, accounting for 65% of the work force. For many of the service industries, recession has hit lightly or not at all, with the result that there has been no good reason for employees to hold back on wage demands or for employers to hold back on price increases—both essential restraints if the rate of inflation is to come down.

The gaining of large wage settlements in the highly visible public and quasi-public service industries—such as health and education—sets a standard, too, that powerful unions in the goods sector have been anxious to imitate, despite recession.

A comparison of this summer's employment numbers with those of a year ago shows that there was a net loss of about 82,000 jobs in manufacturing and a small gain of 12,000 jobs in other goods industries (agriculture, forestry, fishing, mining and construction) for an overall decline of 70,000 jobs in the goods sector (see chart).

At the same time, employment in the service industries in June and July averaged 265,000 above the year-ago figure, with the biggest gain of all taking place in the group of industries known to StatCan by the unwieldy title of "community, business and personal service."

The chances of being touched by recession this year have been especially sizeable for the two groups of people— those who are employed in the goods-producing industries and those newly in the work force and searching for a production job.

This in itself is not a new development. The goods industries have always been the big victim in any recession. The great growth in the service industries through the years means, however, that with each recession, an increasingly bigger proportion of the population has been largely shielded from the direct effect of recession—actual or potential loss of job.

In 1953-54, when the economy was in a recession that was similar in severity to this year's downturn, the goods sector accounted for 53% of all jobs. By the time of the 1960-61 recession, the share of the recession-prone goods sector had fallen to 45%, and when the 1974-75 downturn hit, the share was down to only 35%.

Biggest growth

The biggest growth in the service sector over the years has taken place in public administration and in community, business and personal services, now accounting for 34% of all jobs in the economy vs 28% a decade ago and about 20% two decades back. Looking at the past year, the most remarkable— almost unbelievable—growth in jobs has taken place in "service to business management," a category that includes such activities as computer services, security and investigation services, offices of accountants, advertising services, offices of architects, lawyers, engineering and scientific services, and offices of management consultants.

The sizeable growth in this category is hard to explain. It's possible, although hard to prove, that the slowdown in job creation in public administration— 12,000 net new jobs in the past year vs about 30,000 in 1974 and 1973—has its counterpart in growth of business services through increased use by governments of outside consultants.

All in all, the increasing share of the economy being taken up by services is being looked upon as a mixed blessing. Aside from their immediate recession-proofing effect on price and income decisions mentioned above, they can be indicted, according to some economists, for adding to inflation and maybe even unemployment over the years.

Reasoning

Here's the reasoning behind this view:

• There is generally less scope for productivity advances in the labor-intensive service industries than in the goods sector. The great growth in the service industries vis-à-vis the goods industries has cut into national productivity growth and, in the process, added to the rate of inflation.

• Much of the growth in the service sector's share of the economy has been related to increased emphasis on social goals—expanded government services, for instance, and access to better education and health care. In a detailed analysis of unemployment issues written earlier this summer, Arthur J. R. Smith, president of the Conference Board in Canada, set out these and other national goals that "have come crowding to the fore."

Many, although not all of the goals mentioned by Smith are related to the growth of the service sector, and there is one conclusion that is especially worth noting.

Fundamental choices

According to Smith, "the Canadian people have not yet been effectively confronted with the complex and difficult-to-explain fundamental choices they must face—namely, that strong policy initiatives to secure progress toward these new and highly desirable social goals can only be achieved on the back of substantially higher rates of inflation and unemployment, along with very much slower growth in national productivity and real average personal income growth than could otherwise be attained."

In the short run, any slowing in the creation of new jobs in the service industries would undoubtedly add to the unemployment rate.

But, looking a few years ahead, it seems possible that a turn to a less rapidly growing service sector, especially public and quasi-public services, could well have a beneficial impact from the point of view of inflation and unemployment—if, and it's a big if, the goods industries take up the resulting room given to them.

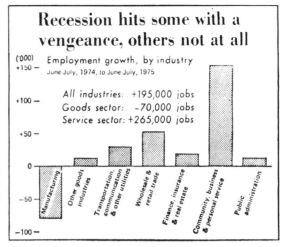

---

1. The GNP figure is always defined in terms of "goods and services" rather than just "goods." How much different do you think the GNP figure would be without the services?

2. In jest it is said that "the difference between a recession and a depression is that when your neighbor is unemployed it is a recession and when you are unemployed it is a depression."

   a) To the economist, what is the difference between a recession and a depression?

   b) Until the Great Depression, Western economic systems had little government activity compared to the Post-WWII period. What do you think was the primary economic reason for increased government activity?

3. Explain in your own words how the growth of the service sector influences inflation. What relation do these arguments bear to the theory of "demand-shift" or "bottleneck" inflation?

4. What has been one major good side effect of the growth of the service sector?

5. The article indicates that the percentage of total jobs in Canada accounted for by the goods sector has declined significantly in the past two years. Yet during this period the goods industry has actually produced more goods. Explain this apparent contradiction.

6. The article notes the remarkable growth in the "services to business management" category. Every time someone is hired for this service category, there are fewer people available to produce actual goods. Does this mean that consumers actually want fewer goods and more of these "services to business management"? Explain.

7. Do you agree with the statement that "there is generally less scope for productivity advances in the labor-intensive service industries"? Discuss.

8. How does cutting into national productivity growth add to the rate of inflation?

9. How would effectively confronting the Canadian people with "the complex and difficult-to-explain fundamental choices they must face" help matters?

10. Explain the logic of the statement in the final paragraph of the article, and clarify the meaning of the "big if."

II-9 Stagflation*

Economists used to think that there existed a trade-off, represented graphically by the Phillips curve (shown by SRP_1 in Figure II-9.1), between inflation and unemployment. This curve is downward-sloping since a rise in unemployment should dampen cost-push forces operating to sustain an inflation, and a fall in unemployment should strengthen these forces. The existence of this trade-off implied that the economy could "buy" a reduction in unemployment with an increase in inflation or "buy" a reduction in inflation with an increase in unemployment. All a policy-maker needed to do under these circumstances was to determine the character of his economy's Phillips curve, choose the point on that curve which was considered the least undesirable, and then adopt monetary or fiscal policy to move the economy to that chosen position. Throughout the 1960s this theory was regarded with some respect; policy-makers eventually subscribed to it and undertook policies accordingly.

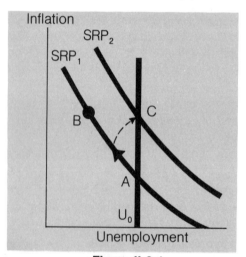

Figure II-9.1

But these policies did not lead to the expected results; if anything things seemed to become worse—the economy began to experience stagflation (high levels of both inflation and unemployment), a phenomenon the Phillips curve theory claimed did not exist. In their search for explanations of this phenomenon economists have considerably modified their conception of the Phillips curve. Regardless of whether or not a stable Phillips curve relationship existed in the late 1950s and early 1960s, everyone recognizes that the world we live in today is quite different, characterized by powerful labor unions, big business, and easily mobilized expectations. In recognition of this new world, the Phillips curve has been modified empirically by incorporating additional explanatory variables and theoretically by developing the concept of the long-run Phillips curve and the natural rate of unemployment.

The long-run Phillips curve is represented graphically in Figure II-9.1 as a vertical line over U_0, the natural rate of unemployment. This unemployment rate is the rate to which the economy gravitates in the long run. It is

---

* This section is for more advanced students.

determined by people's tastes for leisure versus work, by the amount of frictional unemployment (people changing jobs or retraining) required as changing tastes and technology change the nature of what is produced, and by institutional phenomena such as minimum wage legislation and unemployment insurance programs. According to the theory of the long-run Phillips curve, any attempt to push unemployment below this long run natural rate will be successful in the short run, but unsuccessful in the long run.

Graphically, if the economy is at point A in Figure II-9.1, where the short-run Phillips curve SRP_1 cuts the long-run Phillips curve, an attempt by the authorities to stimulate the economy will send the economy up towards B, along SRP_1, increasing inflation and decreasing unemployment. The real reason that producers increase output in response to this policy is that the policy caused prices to rise more than wages, making expansion by firms a profitable venture. After a while, however, labor will realize that prices have increased faster than they had anticipated they would, and will demand catch-up increases in their wages. Furthermore, they will now expect prices to rise at a faster rate and will build this expectation of higher inflation into their future contract demands.

This reaction by labor wipes out the profitability associated with producing at a higher level, so firms contract their output levels, returning unemployment to its original level. But because of the new, higher level of inflation expectations, inflation does not fall back to its original level. On the diagram, the higher expectations of inflation shift SRP_1 to SRP_2 (i.e., at each level of unemployment inflation is higher because of higher expectations of inflation—producers expect to be able to raise prices by more and so agree to the larger wage increases demanded by labor). The economy moves towards point C on the diagram, implying that in the long run the economy moves along the long-run Phillips curve rather than the traditional downward-sloping short-run Phillips curve. This means that there is no trade-off between inflation and unemployment, and that persistent government action to move the economy to a lower level of employment will only serve to push the economy to higher and higher levels of inflation, with no long-run improvement in unemployment.

All of this creates a policy dilemma. Aggregate demand policies (traditional monetary and fiscal policies) cannot improve the unemployment rate in the long run, and can only improve the inflation rate by creating extra unemployment in the short run (to move the economy down the long-run Phillips curve). Wage-price guidelines are often adopted as an alternative policy in this context; their application is supposed to reduce expectations of future inflation and move the economy down the long-run Phillips curve without the painful increase in unemployment so often needed to kill expectations.

Most other policy suggestions in this context involve shifting the long-run Phillips curve (as well as the short-run Phillips curve) to the left by reducing the natural rate of unemployment. These policies are designed to eliminate or alleviate the imperfections in the commodity and labor markets. Examples of such policies are labor-retraining programs, programs to increase employment information or reduce the cost of obtaining such information, programs to increase labor mobility, the elimination of product and labor-market monopolies, elimination of legal minimum-wage rates, restructuring of unemployment insurance programs, and manipulation of government demand and taxation so as to alleviate demand in geographical or industry bottlenecks.

In the first article below the Phillips curve reaches a pinnacle of success of sorts—it appears on the front page! This article recognizes some of the theoretical and empirical developments related to the long-run Phillips curve. Although these developments appeared in the economic literature beginning in 1967-68, and in the textbooks shortly thereafter, not all economic journalists bothered to keep up with these developments. To quote from Keynes (a quote appearing in the second article, which as luck would have it appeared on Friday the thirteenth): "There are not many who are influenced by new theories after they are 25 or 30 years of age, so that the ideas which civil servants and politicians and even agitators apply to current events are not likely to be the newest."

II-9 A Phillips Curve Makes Front Page
Financial Post, September 13, 1975

Stretched-out recovery hurts, but may heal better

By Anne Bower

"WALK, don't run." This is what Bank of Canada Governor Gerald Bouey was really telling Canadian businessmen and consumers last week when he announced that hefty increase in the bank rate to 9% from 8¼%.

Clearly, Ottawa's thinking is that what the economy needs now is not a speedy recovery from recession, but one that is slow, drawn out and, in many respects, painful.

If all goes according to Keynes, the pain, in the form of a prolonged period of high unemployment (currently at 7.3%), should be accompanied by a substantially reduced rate of inflation.

But will it be? Right now, this is a subject of much debate among economists, and they're struggling hard, trying to determine just what relationship now exists between unemployment and inflation.

Some economists believe the clampdown on mone-tary expansion and climbing interest rates so early in the North American recovery could signal the last stomping ground for traditional thinking in economic policy.

As this view has it, if inflation doesn't slow noticeably as a result of the deep recession now over, and the slow recovery now scheduled, the days of the so-called mixed economy as we now know it in North America may be numbered.

Price and wage controls and pervasive planning on a national scale could take the place of today's mix of marketplace decisions and government directives — not because a controlled economy will necessarily work any better, but because our options will have run out. (For what's happening in Britain in this regard, see article p. 19).

To a large extent, it is the changes already wrought in the economy's structure — from fairly perfect competition to fairly imperfect competition — that are raising doubts about whether traditional policies can work to reduce inflation.

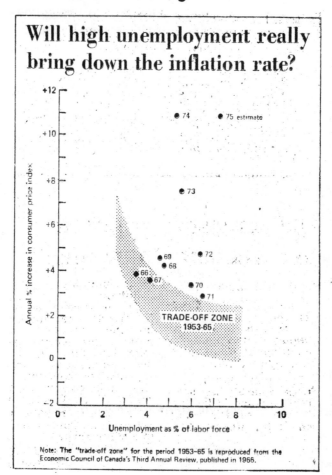

Consider only some features of the economy that have made the 1974-75 recession, and the inflation problem, quite unlike earlier times:

• Built-in price-income rigidities arising from the coexistence of big companies, big unions, and big governments.

• The domination of the economy by the service sector, where the threat of unemployment hardly exists and where productivity gains are slim at any time.

• The massive increase in social benefits that has combined with the growth in two-income families to make unemployment less of a hardship than it used to be.

• Finally, expectations that inflation will continue to be high and must be built into wages in advance.

Despite this greatly changed economic environment, most of the current research into the tradeoff between unemployment and inflation holds out the possibility that the old tools can still be counted on to reduce inflation.

Such a conclusion comes, for instance, from research carried out by Nathan Bossen and Janet Quinnett, economists at **Brault, Guy, O'Brien Inc.**, Montreal brokers. Their analysis of the figures leaves them convinced that "the rate of unemployment is an important variable explaining the rate of inflation and that recent high rates of unemployment promise a significant decline in the rate of inflation over the coming 12-18 months."

The ongoing dissection of the unemployment-inflation relationship by economists everywhere is difficult to summarize, but certain emerging views are worth noting:

▶ Economic life is obviously not as simple as it was in the 1950s and early 1960s, when a fairly stable relationship existed between unemployment and inflation.

In the late 1950s, A. W. Phillips, a British economist, plotted the unemployment rate and the rate of wage increases on a graph and came up with a strong inverse relationship between the two. Thus was born the so-called Phillips Curve, showing rather neatly that when the unemployment rate was relatively high, the rate of inflation (prices or wages) was relatively low, and vice versa.

Curve shifted

In 1966, when the Economic Council of Canada plotted Canadian experience, the main concern was to improve the tradeoff: to move toward a lower level of unemployment at any given level of price increase.

What happened, instead, was that in the late 1960s the tradeoff curve shifted upward (more unemployment for a given level of inflation) rather than downward (see p. 1 chart). Then, in 1973-75, the old tradeoff was wiped completely off the map.

▶ Some recent attempts to redo the Phillips Curve hypothesize that the relationship should be expressed in terms of unemployment and the *change* in the rate of inflation. One such effort along these lines in the U.S. concludes that, viewed in these terms, "there is no evidence the economy has changed in some fundamental way."

According to this analysis of U.S. data by Jack Treynor, editor of the Financial Analysts Journal, there is a significant rule suggested by the data. He postulates that "to reduce inflation by 1%, we must endure unemployment 1% in excess of 'normal' [that is, in excess of 5%] for one year."

A test of the Canadian data by FP did not, however, reveal a particularly neat relationship over the years and, in any event, the 1973-75 experience for both Canada and the U.S. falls outside the unemployment-rate - of - change / inflation hypothesis.

▶ Another school of thought suggests that the unemployment and inflation tradeoff is too simple-minded in any form. According to this view, productivity changes can explain to a large extent why there often appears to be no immediate relationship between unemployment and inflation.

Economists Bossen and Quinnett suggest in an article that "some of the recent tendency for prices to rise even as unemployment should be putting downward pressures on demand is surely due to the productivity factor."

On tight side

As this argument has it, a major easing in the rate of inflation has been hamstrung by a significant decline in productivity over the past year. When productivity "turns favorable, that is when unemployment is high but falling, the rate of inflation could decelerate markedly for something like a year."

Bossen told FP that this analysis of inflation likelihoods assumes that monetary policy in both Canada and the U.S. will be run "on the tight side of neutral." In this situation, he envisages the possibility that the economy could by 1978 have worked its way back to full employment (3.5% unemployment among adult males and just over 5% in total) and to an inflation rate in the 0%-3% range.

Here's hoping.

---

1. What relevance do the ideas appearing in the earlier sections of this book, entitled "Measuring Unemployment" and "Unemployment Insurance and the Work Ethic," have for the interpretation of this article and the policy dilemma it represents?

2. What would a hefty increase in the bank rate cause to happen in the context of the short-run Phillips curve? What would it cause to happen in the context of the long-run Phillips curve?

3. What is a so-called "mixed" economy?

4. What influence, if any, do the four "features of the economy" marked by black dots in the body of the article have on the short-run Phillips curve? The long-run Phillips curve?

5. If "economic life is obviously not as simple as it was in the 1950s and early 1960s," would you be suspicious of any empirical study that did not state whether or not the data used embodied the earlier, allegedly different time period? Why?

6. Why was it that "in the late 1960s the trade-off curve shifted upward... rather than downward"?

7. What theory could lie behind the idea that "the Phillips curve relationship should be expressed in terms of unemployment and the *change* in the rate of inflation"?

8. What theory could lie behind the idea that "to reduce inflation by 1%, we must endure unemployment 1% in excess of 'normal' *for one year*"?

9. Explain in your own words how it is that "productivity changes can explain to a large extent why there often appears to be no immediate relationship between unemployment and inflation."

10. Why should productivity turn "favorable" when "unemployment is high but falling"?

11. In your view, does the graph in the article provide evidence for any one theory of the Phillips curve? Explain your reasoning.

12. The defenders of the long-run Phillips curve theory claim that rather than there being a trade-off between unemployment and inflation, there is a trade-off between unemployment and accelerating inflation. Explain the meaning and rationale of this statement.

II-9 **B Textbook Theory** *Vancouver Sun*, September 13, 1974

But it isn't in the textbooks

From Vancouver's academic world, at the beginning of a new school year, comes word that the most baffled groups of students on campus are those taking economic courses. Much that is in their textbooks has been rendered either out-of-date or downright misleading by the worldwide onset of the Great Inflation. Their economics professors are confused and confusing and don't know the answers.

This state of affairs in our schools and colleges is a reasonably accurate microcosm of what's happening in the larger world. Out there nobody knows how to deal with the contemporary disease attacking the Canadian economy and the economies of all industrialized nations.

The ailment has an appropriately ugly name, "stagflation." That's economists' jargon for a situation which according to the students' textbooks cannot occur: an inflationary recession, when both inflation and unemployment increase at the same time.

No less an authority than Prime Minister Pierre Trudeau has provided a striking example, which no doubt will be preserved for posterity in future schoolbooks, of how wrong are the economic textbooks. After the Second World War it became part of economic gospel that a predictable incidence could be postulated between unemployment and the rate of inflation, showing that the lower the number of jobless the higher the inflationary pressure.

This rubric was cited with particular devoutness at the London School of Economics, one of Mr. Trudeau's old schools. Accepting the textbooks' word, the prime minister apparently decided the Canadian economy stood in as much danger of being hurt by a non-existent creature called stagflation as he himself risked being gored by a unicorn. That led him into the major blunder, after his great triumph at the polls in 1968, of announcing that he would beat inflation by permitting a sharp rise in the number of unemployed.

John Maynard Keynes, high priest of economic theorists, long ago deplored the timelag in the emergence of new economic theories and their adoption by politicians and bureaucrats. "Practical men, who believe themselves to be quite exempt from any intellectual influences, are usually the slaves of some defunct economist ... There are not many who are influenced by new theories after they are 25 or 30 years of age, so that the ideas which civil servants and politicians and even agitators apply to current events are not likely to be the newest."

Keynes' words would seem to hold true today. The politicians and professional economists have not yet updated their thinking far enough to cope with stagflation.

So the layman, like those baffled students on our campuses, can only echo that anguished cry of Casey Stengel whenever he was watching a lousy ball game: "Doesn't anyone here know how this game is played?"

To which, alas, the answer is no.

---

1. Write a brief "letter to the editor" responding to this editorial. Write it as though it were to be published in a newspaper.

II-10 Wage and Price Controls

One of the major suggested remedies for the nation's high level of inflation is wage and price controls. The rationale behind this policy is that much of the inflation is created by cost-push forces: Big business and labor unions are able to raise prices and wages in the face of unemployment because of their powerful position in the economy. This is facilitated by expectations of future inflation. If labor expects prices to rise by 10 percent next year, they will demand a 10 percent wage increase just to keep their real wage from falling. If businessmen expect prices to rise by 10 percent next year, they will expect to be able to raise their own prices by 10 percent and so will be willing to give labor a 10 percent wage increase. Those advocating controls claim that the controls, in conjunction with complementary action on the monetary and fiscal policy fronts to ensure that aggregate demand is not excessive, will reduce inflationary expectations, and thus eliminate a major force in the cost-push process. Critics argue that inevitably the government does not undertake the appropriate complementary monetary and fiscal policies, adopting the controls merely to give the appearance that it is doing something about inflation. Even advocates of controls concede this point: The success of controls depends critically on government adoption of complementary monetary and fiscal policies.

There is, however, another major force in the cost-push process that is more difficult for controls to deal with. In the mid-1970s it seemed that most of these "administered" price and wage increases were sought not so much because of a desire to increase real incomes but because of a desire to attain a "proper" relative income. Prior to the 1970s there was a rough consensus concerning the nature of our income distribution. Spurred by government attempts to redistribute income through a variety of programs, however, almost every group became of the opinion that its relative share of the national income pie was too low. As a result, everyone tried to climb up the income ladder faster than everyone else. But those at the top of the ladder were both strong because they were there and there because they were strong; their fight to protect their invested interests ensured that attempts by others to move up the ladder only resulted in the never-ending scramble called cost-push inflation.

All of this suggests a new kind of trade-off, between inflation and a more equitable distribution of income: As government tries to redistribute income, more inflation is created as the unduly strong use their power to ensure that the redistribution does not take place at their expense. Because it is questionable that controls reduce disagreement concerning how our national income should be divided, opponents of controls argue that what is needed is a policy directed specifically at the powerful in our society.

Although more and more economists are advocating qualified versions of controls as a supplement to appropriate monetary and fiscal policies, the vast majority oppose such controls. The main reason for their opposition stems from their education: As students of economics they are convinced of the benefits to society of efficient allocation and distribution of resources and goods through the price system, benefits that are usually lost under a regime of controls. Advocates of controls meet this point head-on by claiming that the economics profession is too hung-up on the efficiencies of a price system that can no longer legitimately be said to characterize an economy so dominated by powerful private and public organizations. Furthermore, they argue that it may well be the case that the costs of a well-designed controls program (in terms of inefficiencies in the allocation and distribution of goods and services) are outweighed by its benefits (in terms of winding down inflationary expectations without requiring high unemployment). This debate is pursued in the first article below.

II-10 A **A Continuing Debate**   *Financial Post*, October 4, 1975

While the debate in Canada goes on . . .

By Robert Catherwood
and Anne Bower

Will Rogers, the famous American humorist, once suggested a solution to the submarine menace during World War I. He said the thing to do was to get the oceans boiling; this would bring the subs to the surface and the Allied destroyers could then just pick them off at their leisure.

When Rogers was asked how he proposed to get the oceans boiling, he replied:

"Oh, I leave the details to you."

Just as the Allies were agreed on the necessity of eliminating the enemy submarines, everybody today wants to wipe out inflation. It's the details that cause the problem. In a word: how do you do it?

For many, the solution is simple: wage and price controls. Those opposed to controls argue that they have not worked in the past, impose another bureaucracy on the people, and disrupt the traditional economic process to such an extent that the cure is worse than the disease.

One thing is certain: Finance Minister Donald Macdonald won't be lacking for advice in deciding whether or not controls are needed. For example:

Robert René de Cotret, vice-president, Conference Board in Canada, to the annual meeting of the Equipment Lessors Association of Canada:

"One of the most often mentioned non-solutions to our problems is the simple-minded approach of wage and price controls. Let me be quite definite. Wage and price control will not contribute to solving the problems now facing us: it runs the risk of seriously exacerbating those problems and significantly distracting from other more worthwhile avenues of pursuit. To put it very plainly, controls do not work, never have, at best do not achieve their stated goals of reducing price inflation, and at worst lead to severe inequities and misallocation of resources.

No equivocating by de Cotret. On the other side, consider this,

Peter Martin, economist at McLeod, Young, Weir & Co., in a 1975 market letter:

The government must know that it is living in the past, using the same old tools to go after a very different kind of inflation animal — an animal that is just not responsive, at least in the short run, to the laws of supply and demand. Therefore, one has to conclude that Ottawa officials know that the recent bank rate change is not going to reduce inflation, nor is another increase or a third or a fourth. Perhaps inflation can be brought under control with a 12% bank rate and 15% bond yields but clearly this is not a realistic policy. There is but one solution. Let wage and price controls override the temporarily derailed market mechanism for 11½ years. This, would give the adjustment mechanisms time to get back in working order and, more importantly, break the inflation psychology.

Those who support the continued application of monetary measures to contain inflation find a powerful ally in:

Gerald Bouey, Governor of the Bank of Canada in a speech to the annual meeting of the Canadian Chamber of Commerce:

I am aware that many people find it difficult to understand why high interest rates should be regarded as anti-inflationary, since an increase in the cost of borrowing immediately increases the costs of doing business and higher mortgage rates immediately add to the cost of housing. These immediate effects are undeniable, but they are only part of the story and not the most important part.

The most important part of the story is that to react to rising demands for credit by permitting excessive monetary expansion inevitably adds in due course to higher rates of inflation.

Those who advocate controls, however, point to three brief sentences in Bouey's speech, which they claim indicate that the governor does not believe monetary policy can do it alone:

The second proposition is that, whatever else may need to be done to bring inflation under control, it is absolutely essential to keep the rate of monetary expansion within reasonable limits. Any program that did not include this policy would be doomed to failure.
And:
I know, of course, that many things other than monetary policy have also to be right before the economic system will work well."

Does Bouey mean more effective fiscal measures, an income policy perhaps, or both?

Among the most prominent of the critics of monetary policy is the Canadian-born economist:

John Kenneth Galbraith, in a speech given earlier this year and reprinted in the McKinsey Quarterly, published by McKinsey & Co., New York:

The proper lines of action for controlling inflation are not too difficult to envisage. First, we must stop relying on monetary policy to do the job. The monetary tool has been easiest and most convenient for all governments to use because it requires no legislation; it's a matter of men speaking in well-modulated voices around a polished table with charts on the wall. It has been given a good trial. But it hasn't worked, and its economic impact has been discriminatory. Our problem, then, is how to execute a safe retreat from excessive reliance on this instrument which has not served us well in the past five years.

Second...we must recognize that there is no way we can have an anti-inflationary policy that doesn't come to grips with all incomes. There can be no automatic exemption for the incomes of the well-to-do. In an increasingly classless society, with increasingly close comparisons between the incomes received by different social groups, there is just no possibility of getting anywhere with a policy that says: "This is very good for unions, perhaps very good for farmers, but it shouldn't be extended to those in the higher income brackets." Egalitarian taxation of higher incomes is an essential part of any viable anti-inflation strategy.

Third, having committed ourselves to fairness and established a pattern of equality in action, then we must also limit wage claims. There is no escape from some form of incomes policy designed to arrest the wage-price spiral by keeping the claims of the trade union community in a reasonable relationship to gains in productivity.

Fourth, as a corollary to income restraint, I see no alternative to some substantial measure of price control, at least in the U.S. If we are going to restrain wages, we cannot leave profits unrestrained. In a highly concentrated economy such as ours, where the power to raise prices exists and higher prices attract higher wage settlement, some apparatus of control is indispensable, and the sooner we can work one out the better.

Finally, I believe the modern economy requires some new balancing mechanism. In the past we've responded to an excess of demand by shoving up interest rates and restricting credit, with the discriminatory effects already mentioned. If we are not going to rely on monetary mechanisms any more we shall need a substitute and the logical substitute is a much more effective and flexible use of taxation than we've had in the past.

In the U.S. context, this will mean setting our tax levels now to cover revenues at approximately full employment levels, and providing a substantial range on which taxes can be quickly moved up or down as required to balance aggregate demand with supply, assuming that gross imbalances are kept under control by an incomes policy.

Along with academics, industry spokesmen have some strong views on controls. For example:

David Collier, president, General Motors of Canada, in a speech last week:

Wage and price controls would bring unmitigated disaster.

Some of those who believe controls are not the answer have suggested other approaches to coping with inflation:

The economics division of the Canadian Imperial Bank of Commerce in its September, 1975 commercial letter:

The reluctance of governments to rely on the operation of the traditional policy tools has frequently led to a search for

Demands for more by all segments of society are increasing the pressure on Ottawa to institute wage and price controls. Prime Minister Trudeau said last week that controls are not the best remedy." But he kept his options open by adding the time had just not come for controls yet.

new approaches. For example, prices and incomes policies have been tried on several occasions in a number of countries, including Canada and the United States. The U.S. program, which was adopted in the second half of 1971, partly suspended the market system in that country, producing distortion and shortages while suppressing inflation instead of resolving it. Because of the failure of incomes policies wherever they have been tried, more prominence has recently been given to the possibility of broad-based indexing as a solution to the inflationary problem.

There is also a strong body of opinion that says the government should not make any quick moves. In other words, common sense should soon prevail and wage demands, for example, will moderate. But if this doesn't happen soon, controls may be inevitable.

John Parish, senior assistant economic adviser at the Bank of Montreal, in a speech to the Montreal Economics Association:

Obviously, the process of catch-up and leap frogging will eventually have to come under control or prices will just spiral off into the stratosphere. That's hardly in labor's long-run interests and I should think responsible labor leaders are as concerned as anyone over this trend. Given that and all the jawboning that has been going on, the resistance that is being given to high wage demands, particularly by the private sector, and some sobering impact from the high unemployment rate, I think we will begin to get a return to a little realism once the current wave of negotiations is over this fall. If we don't, then look out for guidelines or mandatory wage controls.

---

1. How could controls "run the risk of seriously exacerbating these problems," as de Cotret states?

2. What kinds of "other more worthwhile avenues of pursuit" would Canada be distracted from under controls?

3. How could "an animal" not be "responsive...to the laws of supply and demand," as Martin states?

4. Why does Martin judge a 12 percent bank rate and a 15 percent bond rate to be unrealistic?

5. What is the "inflation psychology" referred to by Martin? Explain its relevance.

6. Interest costs are a cost of doing business. If interest rates rise, costs rise and thus prices rise, exacerbating inflation. Bouey dismisses this argument. Explain more fully his reasons for doing so.

7. What does Galbraith mean when he states that the economic impact of monetary policy has been discriminatory?

8. What is the difference between "suppressing" inflation and "resolving it"?

9. What is meant by a policy of "broad-based indexing as a solution to the inflationary problem"?

10. Which side of this debate do you favor? Defend your view.

II-10 B Old Policies Rejuvenated — Toronto *Globe and Mail*, September 23, 1975

Policies used after war advocated

By EDWARD CLIFFORD

A return to the policies that fuelled Canada's post-Second World War economic recovery has been advocated by the president of the Conference Board in Canada, Arthur J. R. Smith, as a means of pulling the country out of the current recession.

Mr. Smith said Canada allocated a "very large proportion of our productive capabilities" following the war to capital investment in plant, equipment and new housing. It dismantled tight wartime controls, and pursued a course of opening up world trade and restoring trade balances.

In a speech to the National Dairy Council of Canada in Toronto, Mr. Smith said the postwar recovery was engineered by government and business working together. Today that spirit of co-operation is largely gone, however, and both parties must work to find new common ground, he said.

"Otherwise, the cyclical business rampage will continue."

He outlined what he called "seven good reasons" for avoiding wage and price controls in Canada, including:

—No other developed country has been able to make them work over an extended period;

—Controls are difficult to apply;

—Canada is one of the worst countries in which to try to make controls work because of its heavy foreign trade involvement, regional diversification and divided sovereignty;

—They are difficult to operate and absorb too much of the time and energies of people who could be finding other ways of solving economic problems;

—Controls tend to undermine efforts to increase productivity;

—Because many ways can be found to avoid compliance, the perception of inequities causes antagonism;

—Now is the worst time to apply controls because of the large and uneven changes in prices and wages that are occuring.

—Canada finds itself in its present predicament because, through the 1960s, "we had the feeling we had domesticated the wild animal called the business cycle." The nation turned away from productivity growth to a concentration on education, health care, urban renewal, environmental controls and other social measures, he said.

From a level of the second highest standard of living in the world, Canada has fallen to seventh. If present trends are followed, it will have slipped to 17th in the mid-1980s "and still (be) sinking."

Today's problems stem from three causes, Mr. Smith said. Canada is not putting enough emphasis on productivity, it is not encouraging enough new business investment and it is experiencing "an enormous growth in government spending and taxation."

The country must place constraints on social spending, he added, and some of these are beginning to show in a slowdown in education expenditure, postponement of plans for a guaranteed annual wage, reduced funds for housing and slowing of environmental improvement programs.

In advocating a movement away from social programs to those which enhance productivity, he emphasized that he is "talking about economics, not ideology."

---

1. Why should Canada's heavy foreign trade involvement make it difficult for controls to work? Why should its regional diversification do the same? Why should its divided sovereignty also do this?

2. How do controls undermine efforts to increase productivity?

3. Why does the existence of large and uneven changes in prices and wages imply that now is a bad time to apply controls?

4. Why had Canada thought it had domesticated the wild animal called the business cycle? How is such a domestication undertaken?

5. Interpret the three causes of today's problems, outlined in the third-last paragraph, in terms of aggregate supply and demand curves.

6. What is the meaning of the statement that he is "talking about economics, not ideology"? Explain fully.

7. Interpret the suggestion of a slowdown in "social spending" in terms of the trade-off between inflation and income redistribution discussed earlier.

II-11 Interest Rates*

The interest rate is an important economic variable for two major reasons. First, the interest rate plays a prominent role in the operation of the economy and in the structuring of economic theory. It affects aggregate demand, it influences the distribution of that demand, it affects the demand for money, it influences the nation's balance of payments or its exchange rate, it is a determinant of the growth of the nation's capital stock, and it is often used as an indicator of the character of monetary policy, to name a few of its roles. And second, changes in the interest rate are of concern to all segments of society. The investor worries about it because if it changes the value of his bonds will change. The potential home-owner worries because if it changes he could be priced out of the market or be able to buy his home with considerably smaller monthly payments. Importers and exporters worry because interest rate changes can alter the exchange rate and thus affect their profits. The municipal and provincial governments worry because a change may alter the feasibility of borrowing to finance the construction of a new ice rink or hospital. The small businessman is worried about his ability to borrow funds to expand or to ward off creditors. The big businessman worries because it affects the means by which he can most effectively raise capital, through bonds, equity, or retained earnings. The construction worker worries because he knows the strong impact that interest rate changes can have on activity in his industry. The federal government worries because everyone else worries, and because the interest rate level determines the interest payment obligations it incurs when selling government bonds. (The role of the government's financing needs in the operation of the money market and the determination of the interest rate is discussed in the next section of this book.)

Because the impact of interest rate changes is so far-reaching, the financial sections of most newspapers have regular columns updating activity in the money market (a term encompassing the bond market) and commenting on likely future movements of the interest rate. These articles involve considerably more economic terminology than most other articles dealing with economic subject matters, and many involve some technical knowledge of the operation of the money markets. All involve direct application of textbook theory. It is interesting to note the relative lack of discussion of international influences in those articles directed at the U.S. economy's interest rates; the Canadian economy is so open that international influences cannot be ignored.

The factors influencing the interest rate can be categorized in terms of their effects on the supply of and demand for money (the "price" of money being the interest rate; see section I-2C of this book). One major influence is the money supply, for the most part controlled by the Bank of Canada. Open market operations (the buying or selling of government bonds) is the usual vehicle whereby the money supply is changed. For example, a purchase of bonds by the Bank of Canada places money in the hands of the populace (in exchange for bonds), increasing the money supply.

A second major influence is the income level, through the effect it has on the demand for money. With a higher income level people will be undertaking more transactions and will thus require more money to facili-

* Your instructor may wish to cover Section I-2C before moving to this section.

tate those transactions. As a result, the demand for money will increase.

A third major influence is the price level. At a higher price level, a bag of groceries costs more, implying that more cash will be needed to facilitate the same level of transactions. Thus a higher price level implies a higher demand for money. (Often this is explained in terms of the higher price level reducing the *real* supply of money.)

A fourth major influence is expectations of inflation. If prices are expected to rise, money holdings are expected to decrease in value. To avoid this loss people try to get by with lower money holdings. Thus the demand for money falls.

Inflation has an even more important role in affecting the interest rate, which can be more easily analysed through its impact on the bond market. Suppose the interest rate is 5 percent and that suddenly everyone expects prices to rise by 3 percent (instead of zero percent) during the next year. Those loaning money at 5 percent will expect to receive, at the end of the year, dollars that will buy 3 percent less than they would at the beginning of the year, so their expected net return is only 2 percent. To obtain a return of 5 percent they will want to charge 8 percent. Those willing to borrow earlier at 5 percent should now be willing to pay 8 percent because they expect to save 3 percent by buying their car (for example) now rather than next year. In this example, 8 percent is called the "nominal" or "money" rate of interest and is the rate that appears on the market and in the newspapers; 5 percent is the "real" rate of interest, obtained by subtracting the expected rate of inflation from the nominal rate. Because the expected rate of inflation cannot be measured, the real rate of interest is an unmeasurable quantity.

Much of the textbook theory relating to interest rates focuses on the real rate of interest, with the result that students are often unable to relate their textbook economic theory to newspaper reports of interest rate activity. This is easily overcome by incorporating the expected rate of inflation into the analysis of interest rates, translating real rates of interest into nominal rates by adding on the expected rate of inflation. Note that the nominal rate has two basic sources of change: changes in the real rate and changes in the expected rate of inflation. An important implication of this is that real and nominal rates of interest need not always move in the same direction.

A fifth major influence on the interest rate is activity in the international sector of the economy. With fixed exchange rates, balance of payments surpluses or deficits can arise. A surplus (for example) could arise because exports exceed imports or because more capital flows into Canada than out of Canada (a capital inflow usually takes the form of foreigners lending us money). The interest rate is relevant here because capital inflows are a function of the difference between Canada's interest rate and foreign rates. The higher is Canada's interest rate relative to foreign rates, the more capital will flow into Canada, seeking a higher return. (If interest rates in the United States are lower than in Canada, Canadians will be able to borrow more cheaply in the United States and Americans can get a slightly higher interest rate by loaning to Canadians.) A balance of payments surplus means that more money is flowing into Canada during the year than is flowing out, increasing her money supply. This in turn affects the interest rate. A balance of payments deficit has opposite implications. With a flexible exchange rate, surpluses and deficits imply changes in the exchange rate rather than changes in Canada's money supply.

II-11 A Inflation and Interest Rates

Financial Post, September 6, 1975

Saver's dollar may earn more

Resurgence of inflation complicates interest rates

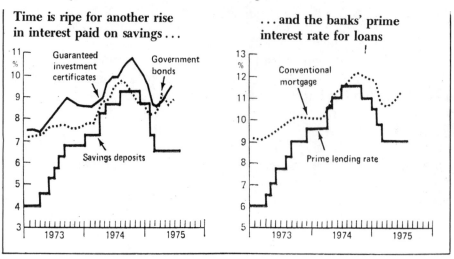

By *Frédéric Wagnière*

REAPPEARANCE of strong inflation both in Canada and the U.S. has led to higher interest rates in all maturities of debt securities.

Rates have gone up on mortgages — and on the guaranteed investment certificates that mortgage lenders sell to raise their funds. Yields have also risen on short- and long-term bonds.

There no longer seems much hope that, by year-end, interest rates will decline to help North American economies in their gradual recovery from the recession.

On the contrary, any sign of economic recovery will only reinforce the trend to higher interest rates again as business and personal demand for credit increases in the face of slow increases in the money supply.

For some months, Bank of Canada has reduced the attention it pays to the levels of interest rates and has kept a close eye on expansion of the money supply.

This policy change has made market interest rates more responsive to the high rate of inflation and to rate increases in the U.S.

Short-term securities issued by lending finance companies and other corporations are yielding about 1½ percentage points more than six months ago, although still considerably lower than last year's record highs.

The central bank's new policy of closely watching growth of money supply has also led to higher treasury bill yields. Large government cash needs and Bank of Canada's reluctance to provide a major part of the federal financing has put considerable pressure on chartered banks as well as other financial institutions to buy government securities, and as a result yields on them are climbing steadily.

Trust companies have reflected the trend in interest rates more faithfully than banks because a greater part of trust company borrowing and lending is on a fixed-term basis. During the last six months, the five-year guaranteed investment certificate rate has risen by about one percentage point to close to 10%. This is still one percentage point lower than last year's high, but all the same it is putting considerable pressure on other interest rates.

---

1. Why should the reappearance of strong inflation lead to higher interest rates? How is this brought about?

2. Explain the reasoning behind the statement that "any sign of economic recovery will only reinforce the trend to higher interest rates again." Is this true of any economic recovery?

3. a) Why would the Bank of Canada have changed its policy to watching the expansion of the money supply instead of watching interest rates?
b) Why should this policy change make "market interest rates more responsive to the high rate of inflation and to rate increases in the U.S."?

4. What are treasury bills? Why would the central bank's new policy lead to higher treasury bill yields?

5. Why should yields on government securities rise if demand for them is higher? Don't yields fall with higher demand?

II-11 B International Influences on Our Interest Rate

Financial Post, June 14, 1975

Interest rates beckon, but U.S.$ stay home

By Frédéric Wagnière

Canadian money-market rates are moving in the opposite direction to U.S. rates.

Investment dealers say **Bank of Canada** is forcing Canadian rates up to shore up the C$, which is just barely holding the US97½c level. This kind of central-bank intervention has to be r e c k o n e d with for some more months — until the economic recovery in the U.S. leads to greater demand for Canadian exports. and the C$ no longer has to be propped up.

In the U.S., **F i r s t National City Bank** lowered its prime rate to 6¼% late last week and most of the banks that had stayed at 7¼% dropped to 7%. This differential of a b o u t two percentage points between Canadian and U.S. administered rates is about equal to the differential in market rates.

This does not mean there is a massive flow of U.S. funds into Canada, because the differentials are mitigated by other factors. In the case of market rates, the discount on the forward C$ has kept step with the interest r a t e differential, and it is only on an unhedged basis that the full advantage of the differential can be gained.

In the case of administered r a t e s such as the banks' prime rate, banking practices are quite different between the two countries. Compensating balances in the U.S. invariably increase the cost of a loan by about 10%-20%. Furthermore, it is not always wise to switch from a Canadian bank to a U.S. bank just because U.S. rates are more favorable. The differential has also been to the advantage of Canadian banks in recent years and switching banks often raises some costs — not the least b e i n g good will.

The forward discount on the C$ and the differences in banking practices between the two countries explain why the interest rate differential does not lead to a huge northbound flow of idle cash.

---

1. How would the Bank of Canada force Canadian rates up?

2. How does this shore up the C$?

3. Why would the Bank of Canada want to shore up the Canadian dollar?

4. Explain how an economic recovery in the United States keeps the C$ up.

5. What are administered rates as compared to market rates?

6. Explain the meaning of the statement "the discount on the forward C$ has kept step with the interest rate differential and it is only on an unhedged basis that the full advantage of the differential can be gained."

*7. What are "compensating balances in the U.S." and how do they increase the cost of a loan?

8. What would be the effect on Canada of a "huge northbound flow of idle cash"?

* For more advanced students.

C The Money Supply and Interest Rates

Financial Post, June 21, 1975

Bank keeps the tap turned to 'cool'

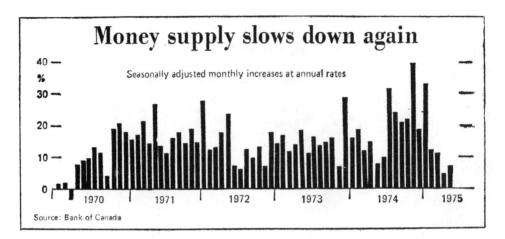

By Frédéric Wagnière

During the past four months, money supply in Canada has been growing at a much slower rate than at any time since last summer.

This slowing down is concrete evidence that the economy is going through a recession and that loan demand is no longer as strong.

It is also an indication that **Bank of Canada's** greater interest in keeping money-supply expansion within reasonable bounds recognizes that the more moderate increases of the past few months can be sustained without exaggerated rises in interest rates.

When Governor Gerald Bouey announced a few weeks ago that the central bank cannot undertake to hold interest rates at particular levels, there was considerable apprehension that rates would rise again.

It is now evident that loan demand has dropped sufficiently since the beginning of the year for money supply to grow at a rate of less than 10% at current levels of interest rates.

Nevertheless, Bank of Canada probably will continue in this direction for some time if only to compensate for the extremely rapid growth in money supply at the end of 1974 and earlier this year— partly reflecting the huge increase in government of Canada deposits.

Interest rates are likely to stay very close to their current levels until autumn if the central bank does not change its policy in the meantime. Rates may then start rising gradually for seasonal reasons. This movement could be even stronger if there are strong signs of economic recovery.

However, forces working in the opposite direction will tend to reduce the upward pressure on rates. World interest rates have not yet stopped declining and the influence of these declines will prevent any excessively sharp rise here.

U.S. interest rates probably have bottomed out at about two percentage points lower than Canadian rates. **First National City Bank** even has a 6¾% prime rate, whereas the Canadian chartered banks are sticking to their 9% prime.

The interest differential between the U.S. and Canada has allowed Canada to cover a large part of its first-quarter current account deficit by attracting funds from the U.S. The unadjusted current account deficit of $2,030 million was financed by inflows of $1,590 million in short-term form and $705 million long-term.

The forward discount on the C$ has meant that there was always a small but steady advantage for U.S. money managers to invest in the Canadian money markets, because the bad Canadian trade performance has weakened the C$. The European countries too, have been lowering their rates and although Canadian capital markets are not generally influenced by rates outside the U.S. and Eurodollar markets, an easing of rates overseas would speed up economic recovery in the U.S. and elsewhere and increase demand for Canadian products.

The economic recovery abroad is the most likely key to a stronger C$, the stabilizing of interest rates, and the possibility of lower interest rates here.

Earlier this week, West Germany's partners in the European Community asked it to embark on more vigorous reflation. Since West Germany is the world's second-largest trading nation, this would undoubtedly influence Canada's exports.

By limiting the expansion of money supply in Canada, the central bank can hope that economic recovery will be the result of strong external demand, and that domestic demand will pick up more slowly and not lead to inflationary pressures.

1. Explain the reasoning behind the statement contained in the second paragraph of the article.

2. The statement "for money supply to grow at a rate of less than 10% at current levels of interest rates" suggests that it is the interest rate that determines the money supply rather than, as the textbooks say, the money supply determining the interest rate. Explain this apparent confusion.

3. Why would a huge increase in Government of Canada deposits be associated with an extremely rapid growth in money supply? Would this necessarily be the case? Explain.

4. Why might rates start gradually rising for seasonal reasons in the autumn? Why are the money supply increases shown in the graph seasonally adjusted?

5. Explain how lower world interest rates prevent an excessively sharp rise in interest rates here.

6. Why are interest rates in the United States lower than those in Canada? Explain.

7. What is "the forward discount on the C$"? What relevance does it have to the "small but steady advantage for U.S. money managers to invest in the Canadian money markets"?

8. Why would an easing of rates overseas speed up economic recovery in the United States and elsewhere?

9. Why is the economic recovery abroad the key to a stronger C$? Why is it the key to the possibility of lower interest rates here?

10. Explain the meaning of the last paragraph.

II-11 **D A Proposal to Curb Inflation** Toronto *Globe and Mail*, January 30, 1976

INTEREST RATE CUT FELT VITAL TO CURB INFLATION

Timothy Pritchard

Policies that would lower interest rates are "the single most important thing government could do to contain inflation on a long-term basis in Canada," according to J. L. Biddell, a member of the federal Anti-Inflation Board.

In a speech to the Empire Club of Canada in Toronto, he suggested that the Bank of Canada take a lead in lower interest rates. He made several other unconventional proposals that, he stressed, were his own and not the board's.

Mr. Biddell said the current wage and price control program will work, but it treats only the symptoms of inflation and offers no long-term cure. Phase two will have to provide for economic growth with less inflation.

"In the absence of other measures, there is no reason to believe that the moment we abandon the present program, inflation will not once again speed up and we shall be back facing recurring bouts of inflation and unemployment."

The key to contining inflation over the long haul is to lower interest rates, he said. Considering the burden high rates create for consumers, businesses and governments. "It seems difficult to justify our central bank's insistence on keeping interest rates at their present level."

Although Canada appears to need foreign capital to pay for numerous projects, it does not make sense to get into a bidding match with U.S. companies and municipalities for money that will be increasingly scarce, he said.

"Obviously, if we are not prepared to pay the price to attract foreign capital, we won't get much of it. If we can't borrow what we need abroad, however, we could do two things.

"First, encourage Canadians to keep their investment funds in Canada by imposing a special tax on interest and dividends received from new foreign investments.

"Secondly, if we can't borrow all of the funds we need from abroad to finance expenditures which we are going to make in Canada in any event, we should at least look at the advisability of the Bank of Canada itself financing some of those expenditures simply by creating the money."

That might be no more inflationary than borrowing the same amount of money abroad to finance the same expenditures, he said.

"I agree that if we curtail our foreign borrowing, the exchange value of the Canadian dollar will decline. This could have some inflationary effect by making it cheaper for other Canadian borrowers to obtain foreign funds."

However, there are various ways governments can "dampen the enthusiasm" of potential borrowers, Mr. Biddell said.

Lowering interest rates in Canada could also be inflationary if credit were made readily available, he added, but there are better ways to control spending than by simply raising interest rates.

---

1. Some people might argue that lower interest rates would be a *result*, not a cause of lower inflation, in direct contrast to the suggestion of Biddell. Evaluate these two contradictory statements.

2. Do you agree that high interest rates create a burden for consumers, business and governments? If so, why do you think the monetary authorities are keeping interest rates so high? If not, why not?

3. What results do you think would stem from a special tax on interest and dividends received from new foreign investments? Explain how this does or does not support Biddell's argument.

4. Biddell's main argument is that if we are going to spend money on Canadian projects anyway, we might as well print the money rather than borrow it from abroad and have to pay interest to foreigners on it. Evaluate this argument.

5. Why would the exchange value of the Canadian dollar decline if Canada curtails its foreign borrowing?

6. Explain the policy implications of the final paragraph.

II-12 Government Financing

There are three basic ways of financing an increase in government spending: collecting taxes, selling bonds, and printing money. If the government raises taxes to finance its extra spending, the higher taxes lower disposable incomes, decrease consumption, and thus largely offset the increase in aggregate demand resulting from the higher government spending. If the government instead sells bonds (this is the financing method usually assumed, when talking of an increase in government spending), the money used to buy these bonds cannot be used to finance other spending, so other spending decreases, again partially offsetting the increase in demand due to the higher government spending. (This can also be explained by appealing to the higher interest rate created by the government bond sales, causing interest-sensitive demand to decline.) When an increase in government spending is financed by the third means (printing money), however, the stimulating effect of higher government spending is aided by the concomitant increase in the money supply.

The first two means of financing an increase in government spending described above involve the "crowding-out" phenomenon: the financing means crowds out other spending, resulting in a smaller policy impact. The monetarists (those feeling that monetary policy is stronger and more reliable than fiscal policy) make much of these crowding-out effects, and claim that the impact of fiscal policy when financed by printing money is due to the increase in the money supply rather than to the increase in government spending.

Whenever the government brings down its budget it announces what its anticipated deficit, or cash requirement, will be. (Of course it could expect a surplus, but in modern times this doesn't seem to happen. The discussion to follow is in terms of a deficit.) The anticipated deficit is the difference between planned government spending and projected tax revenues, a difference that must be financed by the two remaining financing means. A great failure of modern budget presentations is that they do not reveal how much of their deficit is to be financed by selling bonds and how much by printing money. As a result, the degree to which the budget can be considered expansionary (or contractionary, if there is a surplus) cannot be gauged, since the impact on the economy of any change in government spending depends as much on the means used to finance that change as it does on the size of the change.

To the uninitiated, this neglect is compounded by the fact that the government does sell bonds to cover all of its projected deficit. Invariably, however, a large quantity of those bonds are bought by the Bank of Canada; this is what is meant by the expression "printing money." If you and I buy a government bond, our bank balance falls and the government's bank account rises—there is no net change in total deposits in chartered banks and thus their reserves remain unchanged. But if the central bank buys a government bond, the government's deposits in chartered banks increase with no offsetting decrease: the money supply increases.

Because the government has so much financing to arrange, and so frequently has old bond issues coming due that must be refinanced, those predicting the future course of the money markets keep a sharp eye on the government's cash requirements as well as the Bank of Canada's announced policies. It is through the money market that fiscal and monetary policy are most closely tied together; it is when the government needs financing that the cooperation of the monetary authorities is most urgently required. This should become clear from a reading of the articles in this section.

II-12 A The Budget and Financing

Financial Post, May 29, 1976

Canada's lenders will like stand against inflation

By Frederic Wagniere

THE OVERALL THRUST of the budget is just about what the debt markets could hope for. Federal cash requirements are about as high as they were last year and there should be no real difficulty in meeting them.

Of course, if Ottawa needed less cash it would undoubtedly ease the strain on the provinces and the corporate sector.

But as far as the debt markets are concerned, the budget involves no change over what has been happening in recent months.

Aside perhaps from the area of oil/gas exploration, there is no substantial reason for most borrowers or investors to make important changes in their plans.

The budget nevertheless is not unimportant — it stresses two points:

• The goals of fiscal and monetary policy are closely related.

• The deficit will mainly be financed through the sale of securities to the general public and not to the banking system.

For about one year the Bank of Canada has preached and practised monetary restraint. If inflation is to be brought under control, it has argued, money supply should not be allowed to grow at a faster pace than a certain target rate. Although the government never opposed this view, there was no active endorsement of it — until the present budget.

As long as there was no clear indication that the government would not adopt a policy which might have reduced the central bank's efforts to nothing, the idea of investing in debt securities was a gamble on the government's good intentions to fight inflation. The budget makes it clear that the government has taken this task to heart.

It stresses the fact that the deficit will be covered by sales of securities to the general

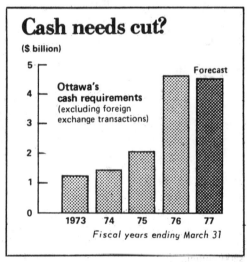

public rather than to the banking system and to the Bank of Canada in particular.

It is a sign that Ottawa has recognized the need to manage its cash requirements within the limitations of a capital market regulated by a well-determined monetary policy. Ottawa seems to avoid the easy way — the seemingly painless way — of financing a deficit: manipulating the banks' liquidity and short-term rates.

What remains to be seen is whether the government will resist the temptation of launching an overwhelmingly attractive Canada Savings Bond campaign this fall. CSBs are indeed a sale of securities to the general public. However, they do not compete with long-term bonds, but with short-term and medium-term bank and trust company deposits. If this year's terms are too good — as they have sometimes been in recent years — the central bank might be forced to take action which would amount to financing part of the deficit itself.

1. Why is it that "if Ottawa needed less cash it would undoubtedly ease the strain on the provinces and the corporate sector"? What determines whether Ottawa or others get the cash?

2. What are the implications of financing the deficit through "the sale of securities to the general public and not to the banking system"?

3. How does the present budget endorse the Bank of Canada's view referred to in the article?

4. Explain the seemingly painless way of financing a deficit, stated as "manipulating the banks' liquidity and short-term rates." Why is it only "seemingly painless"?

5. Explain the logic of the last sentence of the article.

II-12 B How to Raise $½ Billion

Financial Post, July 19, 1975

Ottawa can't afford to make waves in raising $555 million

By Frédéric Wagnière

Ottawa's intentions to raise $555 million in new cash before the end of August seem to be tailored to take advantage of the favorable climate in the money and bond markets while—at the same time—not creating any sudden disturbances.

This extremely largedemand will be done in three main steps:

• Several bond issues will raise $400 million, with Bank of Canada picking up at least $200 million.

• The weekly three and six-month treasury bill tender will continue to be increased by $15 million, raising another $105 million.

• $50 million will be raised Aug. 8 in one-year treasury bills when a former $150 million issue matures.

Investment dealers say that the bond issues are the result of the recent government financing when $165 million was easily raised. This could mean that Ottawa might want to reopen the same issues—9½% bonds maturing in 1994 and 7½% bonds maturing in 1979 and extendible to 1984.

The target would be about $100 million for the long maturity and $300 million for the medium-term one. But, the government may want to try to get as much as $150 million in long-term funds or it may want to add on yet another maturity.

The government and Bank of Canada have been setting the stage for these issues in recent weeks by letting government bond yields rise. Whatever bonds are offered, the rates and the pricing will certainly have to be attractive because chartered banks and trust companies can't easily absorb any more unless their liquidity improves suddenly.

Bank of Canada can also be counted on to take down more of the bonds than the planned $200 million it has announced if they seem to be selling badly.

Indeed, the central bank has surely considered this possibility and has already attempted to assuage fears that there might be a terrific surge in money supply as proceeds from the bond issues are redeposited with chartered banks. It has announced that its purchase of bonds would be offset by a corresponding reduction in the level of its swap agreements with the foreign exchange fund.

In recent years, Bank of Canada has increasingly used these agreements with the government to smooth out short-term variations in the chartered banks' cash reserves. It is also a convenient way to increase money supply gradually without affecting the level of interest rates unduly. Open market operations —the purchase or sale of government securities in the market—tend to have a direct effect on yields that the bank might not think desirable. The purchase by the bank of a large amount of new bonds at the time of the initial offering tends to make money supply advance by leaps and bounds.

One method

One solution to this problem is for the central bank to buy U.S. Treasury securities from the government's exchange fund account. The government's proceeds from the sale are then deposited with the chartered banks and there is a small increase in the banks' cash reserves. The government agrees to buy back the U.S. securities at a set price—hence, these agreements are called swaps.

Bank of Canada has currently nearly $700 million in swap agreements outstanding; it could therefore quite easily buy up the whole issue without adding one penny to the banks' cash reserves.

The size of the issue seems to reflect Bank of Canada's concern over the coming Canada Savings Bond campaign. It wants to reduce the government's reliance on CSBs in order to avoid a repetition of last year' embarrassingly successful campaign. By reducing this reliance, it will be able to fine-tune the CSB rate more easily and avoid an expensive windfall.

Investment dealers agree that this time the central bank will listen more attentively to the business community when it comes to pricing the bonds.

The timing of this issue is also related to the CSB campaign. Usually, new financing is done when old issues fall due and have to be refinanced. There are two issues maturing before the end of the year—$430 million Oct. 1 and $334 million Dec. 15. Both of them are too close to the CSB campaign for large new cash requirements to be raised comfortably, because there would be too many interferences between the two attempts at raising cash.

Nevertheless, the government is expected to raise some new money at the time of the refinancings.

---

1. What are the implications of the Bank of Canada picking up $200 million of the $400 million in bond issues?

2. What does it mean to talk about chartered banks and trust companies "absorbing" bonds? To what does their "liquidity" refer? How could their liquidity "improve"?

3. Why would the government be able to count on the Bank of Canada to take down more than $200 million worth of the bonds if they are selling badly? What does this imply about the Bank of Canada's monetary policy?

4. Why would there be a terrific surge in the money supply if proceeds from the bond issues are redeposited with chartered banks? Why would the proceeds be redeposited with the chartered banks anyway?

*5. The foreign exchange fund is a government agency which holds the country's foreign reserves—foreign currency and securities, for example. Explain how "swap agreements with the foreign exchange fund" can be used to hold the money supply constant, smooth out short-term variations in the chartered banks' cash reserves, or increase the money supply gradually.

6. How can a bond issue be "embarrassingly successful"?

---

* For more advanced students.

U.S. Treasury gobbling up money supply

By Hyman Solomon

WASHINGTON — The soaring demand for funds by the U.S. Treasury is upsetting bond and money markets and raising fresh fears of a credit crunch soon after the economy begins its recovery.

A serious credit squeeze would undoubtedly affect interest rates in Canada and hurt Canadian borrowers in U.S. markets.

"If the bond markets really become congested here, Canadians will still be able to borrow, but they may suffer a little more than domestic borrowers", a New York underwriter specializing in Canadian issues told FP.

"The Canadians might, have to become more innovative and pay more to overcome a sort of nationalistic tendency in such times to look after your own first," he added.

The recent weakening of U.S. bond markets and the sharp increase in money-market yields has been widely attributed to the Treasury's escalating appetite for money to finance the Ford administration's climbing deficits.

Several U.S. companies have already canceled or delayed planned borrowings in the bond markets. The most dramatic example was a decision by **Texaco Inc.** last week to postpone a $300-million debenture issue. Texaco is a prime, triple-A industrial borrower.

"The federal deficit is starting to have its effect," a major bond house official told FP. "Long-term rates have been moving up, and now the squeeze is on the intermediate seven-10-year money. The money might be there, but if the private sector really wants to borrow, it will have to pay."

Short-term interest rates such as Treasury bills also have been moving up sharply.

The British Columbia government has also recently postponed a planned US$125-million issue.

This week, a City of Montreal US$75 million debenture issue was "temporarily delayed" because of unsettled markets.

Treasury department officials have already announced they will need to finance net new borrowings of at least $80,000 million in calendar 1975.

The U.S. deficit for fiscal 1975, ending this June, now is estimated at $45,000 million. For fiscal 1976, the latest estimate is anywhere from $80,000 million to $100,000 million.

The precise size of the 1976 deficit will depend in large part on how expansive the U.S. Congress continues to be on the spending side, and how successful it is in the running battle with President Ford for control of the legislative strings.

Caught in the middle of this struggle is the Federal Reserve Board, which must decide in coming months whether to keep the money hose open to accommodate both private and public demands for credit during a recovery, or whether to allow interest rates to rise in an effort to contain the next round of inflation.

It is a dilemma of major proportions that may well have to be decided sometime this fall, one Fed official told FP.

The federal reserve in the past few months has been expanding the growth of money supply much more than is widely recognized.

An aberration

Although money supply (currency plus demand deposits) dropped in January by nearly 9% (at an annual rate) in what Fed officials consider an aberration, in February it grew at an annual rate of nearly 7%.

And indications are that in March it grew at a 10% annual rate.

This growth has been justified until now because monetary policy has been the major instrument used to help stem the current economic slide and, it's hoped, encourage a turn-around.

With the recent passage of the 1975 tax bill, however, fiscal stimulus should soon begin playing a part in recovery via tax rebates and other measures.

The money markets to date have only had to contend with government needs and with strong demands by corporations for long-term borrowing (mainly to refinance bank loans).

And despite the consternation of the past few weeks on the money markets, there is considerable agreement that the markets can handle the huge federal-government financing needs—until an economic recovery truly gets underway.

But once it does—and the timing is uncertain—there are several strong reasons to fear that private demand for credit will be squeezed, and that some borrowers will come up short or empty-handed.

There is no ignoring the record dimensions of the upcoming deficits and the demands they present to the money markets.

In the last six years, for example, Treasury financing has seldom accounted for more than 20% of the total amount of funds raised in the U.S. economy.

Yet if the Treasury this year is forced to raise $80,000 million or more in net new financing, it could easily increase its share of total U.S. borrowings to between 40% and 50%, a Fed official told FP.

With the Treasury scooping up so much of available borrowings, the main issue will then revolve around the extent of private borrowing needs and the amount of accommodation the federal reserve is willing or forced to provide.

At this point there is no consensus on when precisely the economic recovery will start, or how strong it will be.

Any start to recovery this summer will likely be weak, but this does not preclude the prospect of heavy demand for credit from the private sector.

For example, any modest upturn in the recently depressed housing market could produce a major demand for mortgage credit.

In the same way a turn-around in consumer demand for instalment credit could produce heavy demand for funds. Consumer debt has actually been declining for several months. In February, it showed its first tentative increase, reflecting mainly auto purchases attracted by rebate incentives.

In addition, there are the accelerating demands for borrowed money by state and local governments that have been hard hit by the recession-induced drop in revenues.

Rush for money

And corporations, which have been placing great demands on the bond markets in recent months, can be expected to continue needing money as the recovery gets underway.

Taken together, it adds up to what could be a considerable rush for money in a weak recovery period, some administration officials suggest.

The federal reserve, which may want to rein back on money-supply growth in several months, might find a shrill opposing voice from Congress, insisting that it do nothing to boost interest rates—particularly mortgage rates.

Congress has already been pressing for Fed action to force a lowering of long-term interest rates. The Fed has so far resisted the pressure to relinquish some of its independence, but the recent yield increases on money markets are not likely to help its case any.

"The strain is occurring right now in capital markets," Treasury Secretary William Simon said last week. He said the current level of long-term interest rates at about 9.5% was "extraordinary" in view of the worst recession of the post-war era.

Simon suggested that because of Treasury demands for money, interest rates would not likely drop lower than 8% or 8.5% before turning up again with the recovery.

---

1. Why would soaring demands for funds by the Treasury raise fears of a credit crunch?

2. Why would a serious credit squeeze in the United States "undoubtedly affect interest rates in Canada"?

3. How does the "money market" allocate the available supply of credit to those wanting it? What happens when the government decides it wants more? Explain. What role does the Federal Reserve Board play in all this?

4. Why are the British Columbia government and the City of Montreal issuing bonds in the United States instead of in Canada?

5. Why would a rise in interest rates be interpreted as "an effort to contain the next round of inflation"? What does keeping the "money hose" open have to do with this?

6. How is monetary policy used as the major instrument "to help stem the current economic slide"? How does the 1975 tax bill supplement this?

7. When an economic recovery truly gets under way, government tax revenues should increase, and so should saving (since income will be higher). This should ease the "tightness" on the money market. In light of this, explain the meaning and logic of the two paragraphs beginning "And despite..." and "But once it does....".

8. What is the main issue that revolves around "the extent of private borrowing needs and the amount of accommodation the federal reserve is willing or forced to provide"? Clarify the role and significance of the federal reserve in this matter.

9. Interest rates in the United States (and Canada) have historically been considerably lower than their current levels. This is presumably why the current level of long term rates at 9.5 percent is viewed as being extraordinary, particularly in view of the current recession. Why do you think interest rates are so high in spite of there being a recession, and in spite of the monetary authorities increasing the money supply at a rapid rate?

10. What implications does your answer to the preceding question have for the policy stance of the Federal Reserve Board in the face of "a shrill opposing voice from Congress," noted in the fourth-last paragraph? Explain.

II-13 International Influences*

Canada is an open economy. She is also a small country relative to most other developed economies with which she has economic ties. Having these characteristics means that changes in economic activity elsewhere in the world can have a significant impact on the health of the Canadian economy, and international movements of capital (money) can have a strong influence on the character of Canadian money markets. This should already be evident from previous articles.

Impacts on the Canadian economy from the international sector arise whenever Canada's demand for foreign currency differs from her supply of foreign currency (see section I-2B). Demand for foreign currency arises when Canadians wish to purchase imports or buy financial assets (such as bonds) in foreign countries. Supply of foreign currency arises when we sell our goods abroad (exports) or when foreigners buy our financial assets (loan us money). The main variables affecting these elements of the supply of and demand for foreign currency are as follows. Exports are determined by the prosperity of foreign countries and the level of the exchange rate (in conjunction with the domestic price level). Imports are determined by domestic prosperity along with imports' prices. Capital flows are affected by interest rate differentials, hedging costs, and anticipated future movements in the exchange rate.

The difference between exports and imports of goods is called the balance of trade. When the net value of services sold to or bought from foreigners (transportation costs, for example, or interest payments) is added, this account is called the current account. The difference between foreigners' loans to us (capital inflows) and our loans to foreigners (capital outflows) is called the capital account. The sum of the current and capital accounts is called the balance of payments: the difference between our total supply of foreign currency and our total demand for foreign currency. When supply exceeds demand, Canada has a balance of payments surplus; when demand exceeds supply she has a balance of payments deficit.

The way in which an imbalance in Canada's international payments affects the economy depends on the type of exchange rate system being employed. With a flexible exchange rate system, any imbalance between the supply of and demand for foreign currency is immediately rectified by a "flexing" of the exchange rate, i.e., the "price" of foreign currency. Thus when Canada's supply of foreign currency exceeds its demand, the exchange rate rises, making our imports cheaper and our exports more expensive. This increases imports (increasing the demand for foreign currency) and decreases exports (decreasing the supply of foreign currency), eliminating the excess supply of foreign currency. A similar reaction characterizes a situation of excess demand for foreign currency.

With a fixed exchange rate system, the exchange rate can no longer flex: The government is obliged to exchange currencies at the fixed rate, regardless of whether or not the supply of foreign exchange equals its demand. To see what impact this has on the economy, let us trace through an example of a balance of payments surplus (an excess supply of foreign

---

* Your instructor may wish to cover Section I-2B before moving to this section.

currency). Those holding the excess supply of foreign currency will ask the government to exchange it at the fixed rate. The government usually buys this foreign currency with money it has received from others to whom they have sold foreign currency, but this cannot be done in this case because more foreign currency is being sold to the government than is being bought from it. Therefore those selling the excess supply of foreign currency will be paid for it with new money; a balance of payments surplus will thus cause an increase in the supply of money. Similarly, a balance of payments deficit causes an automatic decrease in the money supply under a fixed exchange rate system. The only circumstance in which this will not occur is when the monetary authorities undertake an offsetting action, "sterilizing" this change in the money supply.

If the monetary authorities do not conduct sterilization operations, this automatic reaction of the economy (changing the money supply) will cause the imbalance in international payments to disappear. For the example of a balance of payments surplus, there is an automatic increase in the money supply, which has the following effects:

a) Canada's income level is stimulated, increasing imports and thereby increasing the demand for foreign currency;

b) Canada's price level should rise, making imports more attractive and exports more expensive. This should increase the demand for foreign currency and decrease its supply;

c) Canada's interest rate should fall, changing the difference between Canada's rates and foreign rates. This should decrease net capital inflows (or increase net capital outflows), decreasing the supply of foreign currency (or increasing the demand for foreign currency).

If the monetary authorities do conduct sterilization operations (i.e., buy or sell bonds to offset any change in the money supply) this automatic adjustment mechanism will not operate, and the imbalance in international payments will continue year after year. A continued deficit will eventually cause Canada to run out of foreign exchange reserves, precipitating an exchange rate crisis and probably a devaluation. A continued surplus causes Canada to gradually accumulate a huge quantity of foreign exchange, angering other countries to the point of precipitating an exchange rate crisis and probably an upward revaluation of the Canadian dollar.

As should be evident from the above discussion, a fixed exchange rate system (or a quasi-fixed exchange rate system in which the exchange rate is allowed to fluctuate only marginally) involves considerable activity on the part of the Bank of Canada. The articles in this section show that monetary policy and interest rates also play predominant roles in the analysis of an economy on a flexible exchange rate system. Thus any analysis of an open economy requires a good understanding of monetary policy and its relationship to the exchange rate; in fact it is often the case that monetary authorities are implicitly given responsibility for the goal of "international balance" in addition to goals relating to inflation and the interest rate. This should be evident from a reading of the articles in this section.

II-13 A Pressures on Exchange Rates

Financial Post, August 9, 1975

Pressure on U.S. rates will persist

By Frédéric Wagnière

The US$ has been exceptionally strong in recent weeks and is starting to reach a level against other major currencies that would represent more closely existing trade patterns.

This recovery has been expected for months because its former depressed level was just too illogical. However, there are various concrete reasons for its firming.

For one thing, short-term interest rates in the U.S. have been going up and there have been inflows of capital from Europe and elsewhere. Furthermore, there is some speculation that the US$ will continue to rise, which has accelerated this flow.

Exports have been extremely strong in recent months and the talk of large grain sales to the Soviet Union has done nothing to damage the overall trade picture.

However, the interest rate differential between the U.S. and other countries is probably the single most important factor. With the U.S. seemingly pulling out of the recession before other major countries, loan demand is going to grow faster than elsewhere, continuing to put pressure on rates.

In Europe and in Japan, the recovery is delayed and the governments will have to adopt slightly easier credit policies in order to stimulate increased economic activity. Rates will therefore continue to decline for some time and it will continue to be attractive for foreigners to invest in short-term U.S. securities.

The strength of the US$ has had but little effect on the C$ because the forward rate has aligned itself on the interest rate differential and there have been very little short-term capital flows between the two countries.

Furthermore a $200 million **Hydro-Québec** issue in New York has boosted demand for Canadian funds. The C$, therefore, has been steady at just under US97c.

---

1. Was the United States on a fixed or a flexible exchange rate system at the time the article was written? How can you tell? What about Canada? Has the United States been recently experiencing balance of payments deficits or surpluses?

2. How does an exchange rate level "represent more closely existing trade patterns"?

3. How does a rise in short-term interest rates affect the exchange rate? Would a rise in long-term interest rates affect the exchange rate in a different way? Explain.

4. Why would speculation that the U.S.$ will continue to rise accelerate inflows of capital from abroad?

5. Why and how would strong exports affect the exchange rate?

6. According to the article, why is the interest rate differential between the United States and other countries changing? What is the economic theory behind how the U.S. rate is rising (and other rates are falling) in this respect?

7. Explain the reasoning (in the next-to-last paragraph) claiming that the U.S.$ strength has not affected the C$.

8. What relevance does a provincial firm such as Hydro-Quebec have to the value of the Canadian dollar? Explain.

II-13 B A Stronger Canadian Dollar

Financial Post, January 24, 1976

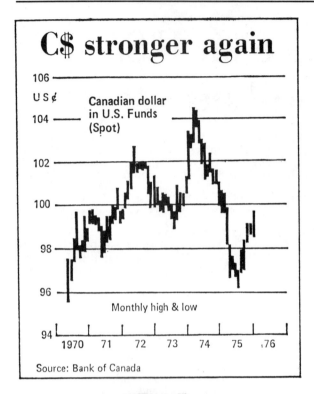

... and that suits our needs

By Frédéric Wagnière

By the end of January, one third of this year's near-record Canadian current-account deficit may already have been financed.

So far, fears that the U.S. and European bond markets might have been saturated by last year's exceptionally heavy Canadian borrowing have not been confirmed.

There may be some signs later in the year that Canadian borrowers are not automatically welcome, but the $1.5-billion headstart means that the prospects are excellent for financing the $4.5-billion deficit without much trouble.

In particular, some large borrowers that are traditionally warmly received in foreign markets, such as the Province of Ontario, have not yet borrowed abroad in 1976. **Export Development Corp.** expects to borrow in Europe for the first time and will surely be well received.

In recent years, short-term capital movements have also financed current-account deficits. For instance, the 1975 deficit — estimated at $5.2 billion — was covered by about $3 billion of long-term capital, $1.7 billion in short-term form, and a $500-million decline in official currency reserves.

Although it is extremely difficult to forecast short-term flows, a net inflow is likely because Canadian interest rates are expected to continue to be higher than U.S. or most European rates (see article on this page).

Despite the large current-account deficit, some upward pressure on the C$ because of long- and short-term capital inflows will make investments in Canada all the more interesting for foreigners. So far this year, the C$ has risen by about 1%, close to par with the US$, because the proceeds of foreign borrowings are being converted into C$ and other conversions are being expected.

This year's improvement of the current-account balance will come from an increase in exports as the U.S. and other major countries pull out of the recession and start buying greater quantities of traditional Canadian exports such as lumber, pulp and paper, ores, and other raw materials. However, the deficit on services will continue to grow rapidly — not the least cause being the increase in debt service costs. Last year's $3 billion long-term borrowing will mean a $300 million increase in interest charges.

Since the beginning of the year, corporations have dominated Canadian borrowing in the Eurodollar market, although Province of Quebec with a $75 million issue did head the list. In the U.S. market, Manitoba Hydro with $125 million and Metropolitan Toronto with $80 million were overshadowed by the announcement that Quebec Hydro had concluded a $1-billion, 25-year private placement at a cost of 10¼%.

This is the third largest borrowing ever made in the U.S. and should cover the utility's capital needs for the whole year — together with $250 million borrowed late last year in anticipation of 1976 requirements. Ontario Hydro is said to be negotiating a private placement of about $300 million.

The outlook for the Canadian balance of payments is therefore encouraging. Foreigners have not been alarmed by the deterioration of the merchandise trade account, which may not be back in the black for years. In fact, strong capital demand and high interest rates could lead to a strong inflow of capital and embarrassing pressure on the C$.

Ottawa might be tempted to increase official currency reserves in order to reduce the pressure on the C$ and the consequent threat to manufacturing industries. But such a move would be difficult. The government would need to borrow large sums on the market because its cash balances are not too high, considering that cash requirements will be close to $6 billion again this year.

1. What is a current-account deficit as distinguished from other kinds of deficits? What would happen if it proved impossible to finance this deficit? Explain.

2. The article suggests there are two major components of the current account, the expected growth in one serving to decrease the deficit and the expected growth in the other serving to increase it. What are these components? Which of these two components is the more volatile? What policy problems does this create?

3. The article lists three basic means whereby the 1975 current account deficit was financed. Which method do you think is the "best" of the three? Defend your answer.

4. If Canada is experiencing such a large current account deficit, why is the Canadian dollar rising?

5. Why should foreigners be alarmed by the deterioration of Canada's merchandise trade account? What would happen if they did become alarmed?

6. Why might Ottawa wish to reduce the pressure on the Canadian dollar? How does it increase its official currency reserves? How would this reduce the pressure?

7. The last paragraph states that increasing official currency reserves would be difficult. Why is this so? Can't the government just print some more money?

8. The article mentions Canada's debt service costs, noting that the *increase* in such costs was $300 million last year. It appears that each year Canadians are paying a huge sum of money to foreigners in the form of interest charges. Wouldn't Canadians be better off it they borrowed this money from themselves so that the interest payments didn't go to foreigners? And if Canadians didn't have enough money to lend, couldn't the government just print some extra money for these purposes, to prevent the huge interest costs from benefiting foreigners? Comment explicitly on the burden of this foreign debt. (Note that much of this debt is undertaken by the provincial governments, so that most taxpayers are implicitly involved in this debt.)

9. The graph shows the Canadian dollar in U.S. funds (spot). What does the "spot" mean?

II-13 C A Too-Strong Canadian Dollar

Financial Post, June 26, 1976

Restraining the C$ may be tough

By Frédéric Wagnière

THE C$ RATE is coming to the forefront of the money markets' concern again. It is now around the US$1.03 level and there will undoubtedly be some pressure on Ottawa from exporters and manufacturers to do something about it.

The upward pressure on the C$ has nothing to do with recent speculation in the foreign exchange markets on the British pound and the Swiss franc. There have been conversions of recent bond issues floated abroad and anticipated conversion of part of the Canadian issues slated for July, which might total $1 billion.

Furthermore, the market is also anticipating the several hundreds of million dollars the Olympic Games will net Canada from foreign visitors. Coming on top of the already considerable capital inflow so far this year, it is not surprising that the exchange rate had to edge up again.

The short-term interest rate spread between Canada and the U.S. has had a tendency to narrow in recent weeks. However the discount on forward C$ has narrowed by the same amount so that there is hardly any movement of short-term capital.

This situation could change rapidly if the C$ were to go any higher, and there could be an outflow of short-term capital. The rising C$ has raised the rate for forward delivery of C$s close to US$1.01 for six and 12 months.

This increase in the forward rate reflects the interest differential between the two countries. However, balance-of-payments projections seem to indicate that, during the next 12 months, the C$ will drop to par or even below. Some importers may already be buying forward US$ outright to cover their future commitments.

If such buying were to develop on a large scale, the forward discount on the C$ might become even wider and short-term investors might then want to invest on a fully hedged basis in the U.S. — buy spot US$, invest in U.S. securities and buy back the proceeds of the investment on the forward market.

Unless there are serious reasons for reviewing the balance of payments outlook, short-term capital movements could put a lid on the C$ rate before it gets out of hand. Of course, if the inflow of long-term capital were to continue at the present rate, the market would be powerless to try to keep the exchange rate down for long.

Official reaction to the exchange rate move could take two forms:

- Controls on long-term borrowing abroad by Canadians.

- Sale of C$ on the market by the exchange fund account to dry up demand for the Canadian currency and a simultaneous decline in interest rates to keep short-term capital effectively out of the country.

Both solutions have drawbacks. Controls on foreign capital inflows might have long-term effects at a time when this capital is needed. Foreign investors are becoming wary of Canada's treatment of foreign capital — present appearances notwithstanding — and controls might be more of a long-term liability than a short-term benefit.

The sale of C$ on the market and a lowering of interest rates in order to manipulate the exchange rate might give undue encouragement to money-supply growth before it is really in the range that would assure price stability.

There might be a little leeway to change rates because the banking system is very tight. However, rates are firming in the U.S. and it is doubtful that Ottawa could have a significant impact on the exchange rate without losing completely control of its basic cautious monetary stance.

---

1. Why would exporters put pressure on Ottawa to do something about the exchange rate?

2. Why would manufacturers do the same?

3. What is "speculation in foreign exchange markets"?

4. What do conversions of bond issues have to do with the exchange rate?

5. What is the discount on forward C$?

6. How does a narrowing of the interest rate spread and a concomitant narrowing of the discount on forward C$ imply hardly any movement of short term capital?

7. Why might some importers "already be buying forward US$ outright to cover their future commitments"?

8. Explain the meaning and logic of the paragraph beginning, "If such buying were to develop..."

9. Does your answer to the preceding question imply that capital would flow out of Canada? If so, how can this be if the interest rate is higher in Canada?

10. How could "short-term capital movements put a lid on the C$ rate"?

11. How would controls on long-term borrowing abroad affect the exchange rate?

12. How would the sale of C$ by the exchange fund account affect the exchange rate? Would this action also affect the money supply? Explain.

13. Explain the connection (hinted at in the article) between the exchange rate and domestic monetary policy. Do any trade-offs exist here?

Should the C$ be devalued?

By Frédéric Wagnière

THE C$ RATE is a little like the weather; many people talk about it, but few do anything about it. Of course, it's not all that easy to do anything about the exchange rate even if you want to.

The government, together with the Bank of Canada, can influence the rate through market intervention or shifts in monetary policy. However, since the C$ was allowed to float freely six years ago, Ottawa has practised a hands-off policy except to smooth day-to-day fluctuations in the market. This view is shared by foreign observers, some of whom might be only too delighted to be able to claim that Canada is a dirty floater.

Nevertheless, as the C$ rose above the US$1.03 level last week, the debate on the exchange rate took on a new sense of urgency. Exporting industries and those that compete with foreign manufacturers in the Canadian market fear that such a high rate will reduce their market shares and cut their profit margins as expressed in C$. Importers, on the other hand, are pleased at this turn of events and claim that cheaper imports are having a mitigating influence on inflation.

No sign of change

Although the central bank has reportedly been selling C$ in the market in recent weeks, there is no evidence that there has been any change of policy in Ottawa. Last week, Bank of Canada Governor Gerald K. Bouey said in a speech in Fredericton: "There is the possibility, in theory at least, that a continuing higher rate of inflation in Canada than in the U.S. could be accommodated by a movement over time of the exchange rate. But it would be a serious error to suppose that such accommodation would work smoothly. One has only to look at the experience of other countries to see how disruptive the movements in exchange rates can be between countries with appreciably different patterns of inflation."

PC spokesmen Alvin Hamilton and James Gillies recently pointed out in the House of Commons that a devaluation is not a solution to the balance-of-payments deficit, because the higher price of imports leads to higher costs in manufacturing, which offset the advantage gained abroad through a cheaper C$.

However, the problem remains: How can Canada find its way out of the present pattern, which leads to current-account deficits of about $5 billion every year?

Gillies takes a medium- or long-term approach to turning around the merchandise trade deficit. He recommends a crash program to develop energy sources in order to make Canada a net exporter of energy again. To increase our exports, he proposes a tax policy "directed at stimulating productivity, at developing our mineral resources, at improving our forest industries and keeping our agricultural sector one of the most competitive and efficient in the world."

The other component of the current account, the service account, is a much tougher problem. The deficit has been growing increasingly rapidly over the past 25 years. From $4.7 billion in 1975, it is expected to rise only moderately to $5 billion this year because of windfall tourism revenues from the Olympic Games. However, the 1977 increase to a forecast $5.8 billion will be more true to form.

The current-account deficit has been financed through long-term capital inflows. However, the wisdom of these inflows and their quality is increasingly being questioned. Ralph Sultan, Royal Bank of Canada's chief economist, told last week's Canadian Financial Conference of the Conference Board in Canada about some of the long-term costs involved in these inflows.

During the first quarter of this year there was a net capital inflow of $2.5 billion — about one half of the projected current-account deficit for the whole year.

"Almost $2 billion of the $2.5 billion was in the form of new bond issues by Canadian governments. At a 9% interest rate, that three months of financial cover will cost Canadians a tidy sum, some $175 million per year in interest payments, year after year, into the foreseeable future. Put another way, that little three months of financial cover will cost Canadians each year about two months of their total lumber exports."

Of course, the first-quarter figures are undoubtedly an aberration and may prove to be the culmintion of a trend. Yet this trend is disquieting — at least as far as the balance of payments is concerned.

During the 1956-60 period, the proportion of Canadian financing done abroad was about 15%; last year it was closer to 30%. For corporations, the proportion dropped to 16% from 28% during those two decades while the provinces increased their foreign borrowings to nearly 50% from 24%.

Latham C. Burns, chairman of Burns Fry Ltd., told the Conference Board in Canada that he now expects a shift back into the domestic markets towards the end of the year. He expects that fully two thirds of total financing requirements for the year have already been fulfilled.

While this may be good news to prospective long-term borrowers, it also might point to one of Canada's weaknesses on the international scene. Latham muses: "One could argue that one effect of the growth in foreign placements has been to maintian the C$ at a high level vis-a-vis the US$; that the offset to any advantage of having junior governments finance massive expenditure programs abroad has been the failure of Canadian manufacturing corporations to become internationally competitive. I do recognize however, that this conjecture is not only very difficult to prove, but also treading on politically treacherous ground."

Sultan believes that the problem of deciding whether we want more larg-scale capital inflows may be taken out of our hands: "In all likelihood the inflow will slow down of its own accord. Provincial governments will begin to encounter debt capacity ceilings — some of them already have. Failing all else, Moody's will accomplish what restless taxpayers cannot."

With foreign sources of capital slowly drying up, Sultan believes that the C$ could be devalued and he says that it is not inconceivable that it could revert to its official IMF parity of US92.5c. He also says that it might be just what the Canadian economy needs and his arguments strangely overlap with Gillies' arguments for no devaluation. In the food and resources area, Gillies calls for a tax policy **to increase productivity and development**. Sultan says that higher profits — the prices being set internationally — would encourage sufficient investment in new and modern capacity to meet rising world demand.

As far as manufacturing industries are concerned, Sultan notes that new investments are increasingly made toward the south because of rising labor costs here. A devaluation might be just enough to restore some of Canada's historical advantage of low labor costs.

The Royal Bank economists have calculated that a 10% devaluation of the C$ would cut the current account deficit in half by 1977; the rate of economic growth would rise to 6% or 7% from a projected 5%; unemployment would drop a full percentage point; and inflation would only be up by one half of one percentage point from their current forecast rate of 7% in 1977.

Although Peter Campbell, vice-president and director of Wood Gundy Ltd., shares the view that capital flows are the main problem, he foresees a different solution. "There has already been a capital inflow of about $5 billion in 1976," he says. "There is another $1 billion under way in offshore markets over the summer, and that still leaves us a lot of year left. Something has to give. Either yield spreads must narrow rapidly or the C$ will zing up even higher. As a best guess, the work-out would appear to be stable or lower interest rates in Canada against some upward movement in foreign markets, with the C$ remaining strong and above par with the US$ for the foreseeable future. This process is already underway."

Provinces major borrowers

Campbell does not believe in action Ottawa could take to limit borrowing abroad. Since the provinces are the major borrowers in the U.S. and in Europe, any attempt to impose some from of controls would either be ineffective if the provinces were exempt, or touch off yet another federal-provincial confrontation if they were not.

If the idea of a C$ devaluation — by whatever means possible — sounds enticing, it seems to some observers that it is not the right time. Past experience has shown that a devaluation is indeed salutary when all other aspects of the economic scene are more or less in tune.

Successful devaluations in the past — such as France in 1969 — were made when wage and prices were higher than in other countries but were otherwise stable or rising at the same rate as elsewhere. Once the terms of trade were changed, distortions across boundaries were of little importance.

Furthermore, a devaluation of the C$ now would have to be forced by Ottawa through some sort of controls on foreign borrowing — a difficult political task. An abrupt easing of monetary policy to lower interest rates might also trigger an outflow of funds that would be sufficient to lower the exchange rate. However, such a move would be completely contrary to the Bank of Canada's stated objectives and would not be consistent with the post-devaluation tightening of fiscal and monetary policy that is always necessary if devaluation profits are to be channeled to the most productive uses.

---

1. Explain how "The government, together with the Bank of Canada, can influence the rate through market intervention or shifts in monetary policy."

2. What is a "dirty floater"? Why might one want to be a dirty floater?

3. Explain in your own words why some sectors of the economy are pleased with the high exchange rate and some are unhappy with it.

4. At the beginning of the fourth paragraph the article states that the central bank may have been selling C$. Why might they be doing this?

5. Explain how, theoretically, "a continuing higher rate of inflation in Canada than in the U.S. could be accommodated by a movement over time of the exchange rate." Would the exchange rate move up or down?

6. What relationship does the "current account" bear to the balance of payments? What are the two components of the current account?

7. What is your personal feeling about Canadians borrowing abroad and incurring future interest payment obligations of the magnitude described by Sultan in the article? Would it be better to borrow domestically, even at a higher rate, on the grounds that we would be paying the interest to ourselves rather than to foreigners?

8. Explain the logic of Latham's comment in the paragraph beginning, "While this may be good news. . . . "

9. Explain the meaning of Sultan's comment that, "Failing all else, Moody's will accomplish what restless taxpayers cannot."

10. Why would the C$ have to be devalued if foreign sources of capital dry up?

11. Explain how Sultan's arguments for a devaluation of the Canadian dollar overlap with Gillies' arguments for no devaluation.

12. Explain the logic of Campbell's statement that "Either yield spreads must narrow rapidly or the C$ will zing up even higher."

13. How would "An abrupt easing of monetary policy . . . lower the exchange rate"?

14. Why would the policy suggested in the preceding question be "completely contrary to the Bank of Canada's stated objectives"?

II-14 Fixed or Floating Rates?*

Textbooks contain lengthy academic arguments about the advantages and disadvantages of fixed and flexible exchange rate systems. In the real world these arguments surface most dramatically at the annual meetings of the International Monetary Fund (IMF), since this body determines which exchange rate system will characterize the world's leading currencies (it should be noted, however, that there often exist maverick members— Canada for many years maintained a floating exchange rate in defiance of the IMF). Crises in the international exchange markets in the early 1970s have highlighted the IMF deliberations in this regard, and have made public discussion and newspaper reporting of this issue more prevalent. A system of floating rates was adopted by the IMF in response to these crises, much to the delight of the economics profession (most economists favor floating rates over fixed rates—economists in general do not like to see the price of anything fixed).

This experiment with floating rates has revealed at least three disadvantages of this system that had not been fully appreciated by economists. The first concerns inflation. An advantage of floating rates is that they can insulate an economy from world inflation. As prices in other countries rise relative to our prices, our exports rise and our imports fall, creating an excess supply of foreign currency. With a floating rate, our exchange rate rises, offsetting the price differential and restoring equilibrium with no domestic inflationary forces created. With a fixed exchange rate, however, the domestic money supply automatically rises (in the absence of sterilization), creating domestic inflationary forces. Although this advantage of floating rates is not usually denied, it is argued that it is gained at the expense of "boxing-in" the inflation in its country of origin. This country's exchange rate is falling, increasing the cost of imports, adding to the domestic inflation, and exacerbating the inflation psychology which is so often the major ingredient of an inflation. This problem is a serious one in countries like Britain and Japan which depend almost entirely on imported industrial raw materials and food, making it very difficult for them to break out of the vicious circle of accelerating inflation.

The second disadvantage of the floating system that was revealed concerns the nature of domestic policy. One of the major advantages of a floating rate is that it frees monetary policy for domestic use (the international sector is automatically looked after by the flexing of the exchange rate). This is cited as an advantage under the assumption that the monetary policy will be used properly. Most democratic governments, however, are naturally prone to pursuing expansionary policies to maintain high employment. With fixed exchange rates there is a built-in constraint—if domestic policy is inappropriate an exchange rate crisis could develop. No such restraint exists with floating rates: Governments can pursue highly expansionary (and inflationary) policies and permit the exchange rate to fall without incurring the loss of prestige (and rates) associated with a formal devaluation. The result, with everyone pursuing a similar policy, is a framework conducive to international inflationary pressures. Many would argue

---

* The introduction to the preceding section should be read before moving to this section.

that the world-wide inflation that began in 1973 was triggered by the floating of the U.S. dollar, and thus the entire exchange rate system, in August of 1971.

The third disadvantage concerns the volatility of exchange rates and the resulting reduction in trade because of uncertainties. Canada's successful experience over the years with a floating exchange rate was often used as an example to support arguments that a floating system would not be characterized by volatile rates. Unfortunately, we have now learned that the market for the Canadian dollar is not typical. It is a very efficient market, with many transactors and many transactions: it is a *broad* foreign exchange market. Most other foreign exchange markets, it has turned out, are *thin*: there are few transactors and few transactions, causing the exchange rate to jump around daily. The resulting uncertainty causes hedging costs (the cost of insuring oneself against a change in the exchange rate) to be high. This increases the cost of foreign trade, reducing trade and the benefits it provides.

The two articles in this section reflect some of these feelings concerning the undesirability of floating exchange rates, but at the same time defend their good points.

II-14 A Floating Rate Suits Canada

Financial Post, June 14, 1975

Floating C$ suits us, other currencies seeking anchor

By Frédéric Wagnière

Most countries of the world are slowly coming to the conclusion that they are not prepared to live in a world of floating exchange rates — after having been forced to put up with them for some two years.

Ironically, the trend toward fixed exchange rates, comes just after floating rates eased the recycling of surplus Opec revenues back to the importing countries.

Nobody is advocating a restoration of the Bretton Woods system, which broke down four years ago. But many countries are disillusioned by the results of floating and are trying to tie their currencies to those of their major trading partners.

There is no such move afoot in Canada, where the floating of the C$ in May, 1970, has generally been viewed as a success; the only complaints come from those who say the C$ rate is either too low or too high.

Low cost

The main reason that a floating C$ works in Canada is that a vast majority of trade and investment transactions are with the U.S. An efficient foreign-exchange market has developed between the two countries. The cost of hedging future transactions is seldom higher than 2% a year, because of the broad forward-exchange market.

Other countries are less fortunate.

The system of generalized floating rates was reluctantly adopted slightly more than two years ago, because huge speculative movements of capital were triggering currency crises more and more frequently. Floating rates, it was argued, would pit the speculators against each other and spare the central banks the troubles they had gone through so often in the past.

This aspect of floating rates worked well enough. Some of the banks that continued to speculate in the foreign-exchange markets ran up huge losses and went bankrupt — a sobering thought for the other banks.

Negative effect

But some of the disadvantages of the floating-exchange rate system have become apparent.

The markets have seen a decline in the volume of transactions and have become much more volatile.

As a result, hedging costs have risen sharply — an obvious impediment to orderly international trade. For instance, between Germany and Switzerland on one hand and Britain, France, and Italy on the other, hedging costs have generally been between 5%-10% a year in recent months.

Moreover, floating rates should normally have a negative effect on the exports of the strong currency countries and boost the exports of the weak countries, because as a currency goes up the price of exports expressed in other currencies also goes up.

However, if this effect is nearly automatic for natural-resource products where an international market exists, it takes a much longer time to make itself felt in the fields of consumer and capital goods where substitution is not so easy.

Although the strong currency countries can expect to see their competitive position fade in coming years, they are also enjoying an important advantage in a period of high inflation. The goods they import become cheaper, because one unit of their currency will buy more abroad.

This was of special importance at a time when the price of oil was quadrupled. Although the main impact was only to slow down inflation, in some cases there were price cuts. For instance, some months ago, **Ford Motor Co.** was able to cut prices on all imports into Switzerland by 10%.

By contrast, weak currency countries such as Britain and Italy have had the worst rates of inflation in Europe.

There was also some hope that the floating system of exchange rates would eventually lead to a system of stable rates. The relative stability of the C$ — which has always been within 4% of par with the US$ during the past five years — has not been duplicated by other currencies. Whereas the C$ rate generally changes by less than half of a percentage point a day, other rates often change by more than one percentage point in one day because the foreign exchange markets are so thin. Under these circumstances, it is small wonder that many countries are finding it advisable to enter into a system of fixed exchange rates, even if this system is not universal.

France is scheduled to rejoin the European snake (the system whereby Germany, the Netherlands, Belgium, Denmark and, to a lesser degree, Norway, Sweden, and Austria agree to keep their currencies within a 2½% margin).

Switzerland is also planning to become an associate member and Italy is hoping to get its finances in shape in order to take its place alongside its European Community partners. Japan may also become the hub of a regional system of fixed rates.

In the meantime, the Opec countries have grown impatient with the volatility of the US$ and they decided earlier this week to switch their pricing mechanism to special drawing rights (or SDR, the unit of account of the International Monetary Fund) from the US$. The SDR is the weighted average of 16 major currencies and is, therefore, less prone to sudden fluctuations.

So-called baskets of currencies have been in use for a long time inside the EC and in the Eurodollar market, as they are a convenient way for creditors and debtors to avoid any sudden and excessive fluctuations.

The SDR may turn out to be the most important of these international units of account because it comes closest to being universal. It has now made its way into private finance after being considered as the exclusive preserve of central bankers and governments. Recently, **Swiss Aluminium Ltd.** issued a Eurocurrency issue of 30 million SDRs (US$37 million). Other borrowers may follow suit.

---

1. How would floating rates ease "the recycling of surplus Opec revenues back to the importing countries"?

2. What is "the Bretton Woods system"?

3. Who complains of the C$ being too low or too high? Why?

4. What is "the cost of hedging"?

5. Why does the fact that there exists a broad forward-exchange market keep the cost of hedging as low as 2 percent? Is this explanation the same as the explanation given for the high hedging costs in Europe? Why or why not? Explain.

6. What are "speculative movements of capital"? How do they trigger currency crises? How would floating rates pit speculators against each other?

7. What is the basic difference between natural resource products and consumer and capital goods that makes substitution not so easy? What has this to do with elasticities? Explain.

8. Explain how Ford was able to cut prices on imports into Switzerland by 10 percent.

9. The article's statement "By contrast, weak currency countries such as Britain and Italy have had the worst rates of inflation in Europe" is not clear about whether a weak currency augments inflation or vice versa. Which theory do you support? Why?

10. Does the existence of the European snake increase or reduce hedging costs from what they would be in its absence? Explain.

11. Could the so-called baskets of currencies or SDRs be used to avoid the high hedging costs? If so, how? If not, why not?

II-14 B An Exchange Fight

Vancouver Sun, August 20, 1975

U.S. prepares for exchange fight

By KEVIN DOYLE

WASHINGTON (CP) — United States treasury officials have begun quietly bracing for an anticipated major push by France later this year to force a return to a fixed set of exchange rates for the world's leading currencies.

In fact, some officials say they expect the issue may become an important talking point as early as the annual meeting of the International Monetary Fund IMF here early next month.

France still appears relatively isolated in its desire to abandon the present international system which allows currencies to fluctuate widely in response to world demand and supply. But it has recently been giving strong indications, officials here say, of getting ready to gear up its campaign.

The French motives for wanting a return to the fixed-rate system of currency values which prevailed before the early 1970's are partly self-interested, partly political and partly economic. But it is on the economic issues that France will have to make its case if it is to have any chance of getting serious support from other countries. And while these concerns alone would fill volumes, the French are expected to concentrate largely on the one with the most popluar appeal: They contend that floating rates are a major cause of world inflation.

The French argue that when a major currency — say the United States dollar — depreciates, it makes the price of imports to the U.S. higher. This means that consumers and companies are left with less to spend on domestic goods, the economy goes into a slump, the jobless rate rises and pressure on government to take action grows.

The action too often taken, the French contend, is to increase the domestic money supply by amounts so large that demand for domestic goods to substitute for expensive imports soars beyond the capacity of the economy to meet it and inflation becomes inevitable.

But a large currency depreciation under a floating system, the French say, also has other inflationary effects. When the dollar, for example, decreases in value against other currencies it makes U.S. exports progressively cheaper to overseas buyers and as their demand grows, domestic producers can shove up prices without much fear of losing their foreign customers.

If the same goods are sold domestically U.S. buyers then find they, too, are facing a situation where the dollar buys less of the product than before, and thus, say the French, depreciation has set in motion a series of events which raise the general level of U.S. prices and increase the pressure for higher wage awards, causing even worse inflation.

The over-all result when this happens in several big countries at the same time, French officials say, is a gradual worldwide spread of inflation.

The U.S. and many other countries, such as Canada, which favor floating rates have a compelling range of aruguments to counter the French position.

From a U.S. point of view, of course, a depreciating dollar has the obvious attraction of having been largely responsible for stimulating exports and decreasing imports during the last couple of years and turning a disastrous trade deficit into near balance.

But the broader argument levelled against a return to fixed rates is that it would again lead to a constant series of monetary crises with speculators attacking weak currencies— such as the British pound — and central banks being forced into huge support operations to maintain par values.

And while floating currencies may hurt consumers somewhat by raising prices slightly, U.S. officials argue, they have also made it possible for many countries to create jobs by increasing exports and in some cases to save the export industries involved from bankruptcy.

With the huge trade deficits created in many Western countries by the quadrupling of world oil prices, these experts say, floating rates and constant adjustments have been a major factor in averting a complete breakdown in international trade and monetary systems which might have created a depression on the scale of that of the 1930s.

But officials here are the first to acknowledge that however strong they believe their position to be, the next few months are likely to see their arguments tested to the full.

Macroeconomics 129

1. In the fifth paragraph of the article, what does the French argument assume about the price elasticity of the U.S. demand for imports? Explain.

2. Also in the fifth paragraph of the article, what does the French argument assume about the price elasticity of foreign demand for U.S. exports? Explain.

3. Does the second sentence of the seventh paragraph seem to be self-contradictory? Explain.

4. In the eighth paragraph, what is the meaning and logic of "U.S. buyers then find they, too, are facing a situation where the dollar buys less of the product than before"?

5. How could a flexible exchange rate system have turned a disastrous trade deficit into near balance? What would have happened had the country employed a fixed exchange rate system?

6. What is a "weak" currency? Why and how do speculators "attack" weak currencies?

7. Explain how the example of the quadrupling of world oil prices provides an example of the advantage of a floating rate system.

8. Many new economic theories reach conclusions and recommendations counter to traditional economic thought by incorporating into their analysis different assumptions about the economy's reactions to situations in which "demand does not equal supply." One common specification in this regard is overreactions on the part of policy makers. Comment on the role this assumption plays in the French argument against flexible exchange rates.

II-15 Growth

Growth theorizing in the post-WWII era has for the most part been highly mathematical in nature, with little thought to the relevance of the theory to the problems of the real world. In fact, until the 1960s, most growth theorists blithely assumed that growth was in and of itself a good thing, and they completely ignored any possible side effects that growth might have (such as environmental damage). Eventually reality forced economists to realize that perhaps growth might not be an unqualified good thing. Pressure from zero-growth advocates has forced the majority of the profession to espouse "growth within constraints" as their growth goal. Growth is required to provide jobs and maintain living standards for a growing labor force, and to make redistribution easier by creating a larger output pie. The constraints are required to avoid environmental damage and to ensure that the growth occurs in the "right" sectors of the economy.

Capitalism, if left to itself, will not necessarily produce "growth within constraints." As a result, much attention is directed to the question of how government might control the growth dimension of the economy. The two articles in this section illustrate the basic nature of this growth controversy. Much of the debate is subjective and many of the policies suggested by the debate are of questionable value, but there is no doubt that society must continue to search for answers to the problems raised by this controversy.

Major capital shortage

THE GREATEST, long-term problem facing the American economy is the rapidly developing and little appreciated capital shortage.

The chairman of the New York Stock Exchange, Mr James Needham, has been speaking on this point for the past year, arguing that the shortfall will be about $650,000 millions during the next 10 years.

Others have taken up the cry. Recently, Business Week magazine noted that the United States used $760,000 millions in new capital from 1955 to 1964, and $1,600,000 millions in new capital from 1955 to 1964, and $1,600,000 millions in the next 10 years. The magazine projects the requirement for 1975-1984 to be about $4,500,000 millions.

Because of it, the editors write, the American economy may be on the verge of drastic alterations, "marked by slower growth, higher unemployment, and fewer promises for nearly everyone." In effect, the United States retains its high standard of living, but the result will be slow decay and ultimate paralysis, something Britain has experienced for the past generation. Or it can deliberately lower consumption to rebuild its capital plant. The payoff would be a stronger economy for future generations to build upon. This is what the Germans and Japanese did in the 1950s and 1960s.

The American economy is capable of producing a finite amount of goods. During the past generation and more, it has concentrated on consumer items to the detriment of capital goods. As a result, America's industrial plant is old, and in some cases antiquated in comparison with those of some other nations — again, Germany and Japan are the prime examples here.

What Mr Needham, Business Week, and others like them have been saying is that the consumption boom has to come to an end, that Americans must be prepared — as a nation and individually — to consume less and invest more, if the country is to avoid the British disease.

This does not necessarily mean Americans must have a drastically lower standard of living. A largescale redistribution of wealth might go far toward accomplishing the task, as could a major leap in worker productivity.

The Government might channel funds for increased investments through a new version of the Reconstruction Finance Corporation (RFC), which would lend money at low rates for new plants, or modernisation of existing plants and equipment. But simply printing additional money would only result in still greater inflatiuon.

Even without such an agency, the Government will show a deficit of more than $60,000 millions during the next 12 months. This will be financed through sales of Government paper, which will increase competition in the capital markets and result in higher interest rates, which, in turn, will mean higher prices — inflation — which usually feeds upon itself. An alternative would be heavy taxes on consumer items, thus cutting down on expenditures and raising money for the new RFC. In this way, the Government would act as a capital pump and a vehicle for redistributing wealth and directing the economy.

There is another way to accomplish the task, one that the Federal Reserve chairman, Dr Burns, and the Treasury Secretary, Mr Simon, have alluded to, though not directly. Both men would like to reduce Federal spending, obtain a balanced Budget, and do all in their power to ease pressures on the capital markets.

Their unspoken programme centres on much higher corporate profit. In their view, these are not "obscene" at present, but in fact far too low. The energy industry alone will require about $900,000 millions for capital and related spending in the next 10 years.

This money could be raised by a combination of profit, borrowing, and the flotation of new common stock issues. If profit is high, the companies would have less reason to go to the capital markets, and when they did, interest rates would be low if they borrowed to buy new equipment, or to build new facilities. In addition, their high-prices common stock would be easy to sell. Such an idea is understandable, but imposible to sell politically, at least in this form. Most middle-income Americans are suffering and believe corporate profit is too high as it is. Can they be expected to accept a plan which, in effect, increases profit and further lowers their standard of living?

A new bull market would be most helpful, for then corporations could float additional common stock issues and so obtain funds that way. If present trends continue, only the largest, the most trusted firms will be able to sell equity, while small, and medium-sized companies will be starved and forced to decline, go out of business, or be taken over by the giants.

Whatever means are used, Americans will have to consume less, and produce and invest more, both as individuals and as a nation. Business Week estimates that agriculture will require $400,000 millions in the next 10 years for capital improvements; building $1,900,000 millions and steel $50,000 millions to accomplish their tasks. In effect, the steel worker will have to lower his standard of living and turn out aditional products so that the company will have a higher profit, which will be ploughed back into operations and so create more jobs and more steel. Not to accept this concept could mean slow economic strangulation. — Los Angeles Times-Washington Post.

Robert Sobel

1. Why does the article cite numbers in millions when it would appear easier to cite them in billions? (Don't waste time on this question if you don't know the answer.)

2. Why would a shortage of capital slow growth and lead to higher unemployment?

3. How does an economy "deliberately lower consumption"?

4. Why would we want a stronger economy for future generations to build on? What did future generations ever do for us?

5. What is the theory behind the statement that "higher interest rates, which in turn will mean higher prices—inflation—which usually feeds upon itself"? What is your evaluation of this theory?

6. How does reducing Federal spending and obtaining a balanced budget ease pressure on the capital market?

7. What is your opinion of the plan to solve the capital shortage problem by raising corporate profits? Why would it imply low interest rates?

8. The article stresses the need to reduce consumption and increase investment (capital formation). The capitalist system uses the interest rate as an allocating mechanism. But a higher interest rate, although reducing consumption, will reduce investment; and a lower interest rate, although increasing investment, will increase consumption. Is there something wrong with the capitalist system here? Explain, and comment in the light of the contents of the article.

II-15 **B A Conserver Society** *Toronto Star*, March 6, 1975

Ottawa's new dream: a 'conserver society'

Robert Nielsen
An article from the Toronto Star.

The Trudeau government is contemplating a profound change in the Canadian way of life, from an economy based on growth and rising consumption to a "conserver society."

In fact, the government has already gone beyond contemplation to advocacy and preparation. While the prime minister and Energy Minister Donald Macdonald drop prophetic hints of the new day to come, scholars are spending fair-sized chunks of federal money to plan for it.

Nearly one-third of next year's $1.8 million budget of the Science Council of Canada is earmarked for this purpose, and another $195,000 will go to an "associated future studies group" at the universities of McGill and Montreal. No fewer than eight departments or branches of the government are supporting the project.

The philosophical warrant for all this activity was written in a little-noticed speech by Trudeau in 1971:

"Suddenly, in the past two decades, the rush of technology has become so swift, the siren song of material gain so seductive, that we have permitted commercial processes to commence that we do not always understand, and which may lead to disastrous consequences . . .

"We haven't demanded that each new innovative process be accompanied by a critical process which is prepared to analyze it.

"No businessman would calculate his net gain without first taking into effect the deterioration of his machinery and the depletion of his stock of raw materials."

Why then, the prime minister asked, do Western governments "continue to worship at the temple of the gross national product" without taking into account resource depletion as well as harmful environmental and social effects?

The phrase "conserver society" first appeared in a Science Council report of January, 1973, in the high context of a global mission. The council recommended that Canada "begin the transition from a consumer society preoccupied with resource exploitation to a conserver society engaged in more constructive endeavors." Ideally, Canada could provide the leadership "toward more equitable distribution of the benefits of natural resources to all mankind."

This embryo of an idea is now starting to sprout a full set of tissues, organs and

Macroeconomics 133

limbs — or rather, several sets of them; alternative models of a new society which will be markedly or even radically different from the one we know. The first tentative blueprints are expected to be ready by mid-year.

These futurist researches take off from two premises:

(1) That material economic growth can't go on indefinitely, as it has done since the beginning of the industrial revolution; it will be stopped either by exhaustion of non-renewable resources or by incapacity to develop them safely.

(2) That economic growth raises the quality of human life only to a certain point, and then proceeds to lower it.

On the first point Dr. Ray Jackson, a physicist with the Science Council, doesn't claim that the world faces any absolute scarcity of resources; he allows for the possibility that the earth contains a million times the quantities of minerals and oil that have so far been discovered. But he believes that the energy required to get at them could generate a calamitous amount of heat. An average temperature rise of less than one degree Celsius would, he says, melt the polar icecaps.

Jackson and Dr. Arthur Cordell, a Science Council economist, are passionately opposed to continued economic growth on the familiar scale and pattern, and to what they see as a mindless addiction to growth in both communist and capitalist countries.

As an example of growth leading to degradation, Cordell offers New York City: "It has the highest per capita income of any American city, but who would say its quality of life is the highest in North America?"

Cordell believes we are suffering from a "mind-set" and institutions that got started during an age of scarcity and are now obsolete. We have solved the problems of agricultural and manufacturing production, but we just go on doing more of the same without asking why, or counting the costs. "Busyness" has taken on a life of its own.

The automobile industry strives for endlessly rising production and sales of cars, not to provide the public with optimum transportation, but on the assumption that more is better (at least for the auto comanies).

Auto workers besiege governments for action to end the slump in the industry, not because the public needs more cars — if it did, it would presumably be buying them — but because they consider jobs essential to their social prestige.

If the gross national product is equated with the quality of life, then, Cordell says, the logical thing would be to have two model changes per year in the auto industry — "more employment, and a rising GNP."

Jackson and Cordell would like to design a substitute for the GNP as an indicator of a society's well-being. An economic activity wouldn't get a plus mark just because it produced goods and paid wages; such costs as clearing up any pollution it caused, or any extra police it made necessary, would have to be deducted from the value of the benefits.

In this way they would try to compute a "net national product," but it's hard to make calculations which will be generally accepted as unbiased. Because of this difficulty, they have virtually abandoned hope of coming up with a single number to express the concept.

In the conserver society, growth would cease to be an end in itself and would become merely a byproduct of what people chose to do; but choice in economic activity would be limited by tests of social usefulness of acceptability.

Techniques that promote high levels of consumption — such as advertising and credit buying — would be questioned. Design criteria might require goods to be durable, repairable and recyclable — "built-in obsolescence" would be out. So would useless proliferation of nearly identical consumer products; tax deductions might be disallowed on the development costs of a product unless it represented a real innovation.

Cordell and Jackson don't deny that this would stifle some kinds of initiative and enterprise, but they foresee clear gains, not only in better allocation and husbanding of resources, but also in people's satisfaction from work. Much present work is "alienating," they contend, because it is not related to real human or community needs.

"We need only one-fourth of the working population to produce manufactured goods, and much of what they produce is junk," says Jackson.

"The system is perverse," adds Cordell. "It demands a whole lot of people who consume but don't produce. Is the U.S. economy in trouble? 'Go out and buy!'

"The sad part is that as the system matures, we have to keep hyping it up all the time or it'll collapse on us."

How, then, would the conserver society meet people's demands for jobs and incomes?

"You have to look at the problem in a totally different way," says Cordell. "One way of breaking this damn thing — the work-cash nexus — is shown by the Opportunities for Youth and Local Initiatives Programs. That gets money into people's hands for community work they choose themselves."

But aren't such programs only made possible by the surplus wealth which a growth economy produces?

"All that productivity has done is freed time — allowed people time to do other things like OFY and LIP," replies Jackson. "We confuse ourselves if we say we can only do these things because of the creation of wealth."

The conserver society will require a transformation of attitudes and institutions, according to these Science Council advisers.

E x a m p l e s : Competitiveness would yield to a co-operative spirit; education would instill "a sense of community contribution" instead of being "a job and consumer-training system."

For those who see in all this a bleak prospect of reduced living standards and restricted choice, Cordell has an upbeat answer.

"The whole notion of a conserver society may be one of the most liberating things in the history of mankind. It could release millions of people from producing junk and pushing paper just so they can have jobs and incomes. The potential for increased leisure is enormous."

The models for such a society will range from the "Scottish," concentrating on input and efficiency, to the "Buddhist," entailing a cut-down in consumption, according to Prof. Peter Sindell, a McGill anthropologist who is assistant director of the McGill-Montreal project

The researchers agree that, whatever its form, the conserver society will have to come to grips with the politically explosive matter of income redistribution. If the total national pie stopped growing, or grew more slowly, it would in fairness have to be more evenly shared.

Professor Dusan Pokorny, a University of Toronto economist, has drawn part of this assignment, and he is aware of having a hot item.

In one sense Canada is well situated to distribute income more evenly, Prof. Pokorny says, because of its advanced development and abundant raw materials.

"But as far as social psychology is concerned, this country is far from ready." Canadians base their expectations on continued growth, and it would be a tremendous wrench for them to adjust to economic "equilibrium or stagnation."

Social conflicts are avoided by rising growth because it enables many people to realize their expectations. If growth stopped, Prof. Pokorny thinks this would generate social tensions too acute to be resolved by "traditional means."

At this preliminary stage of his inquiries, he is thinking not of zero overall growth, but of variable rates of growth for different parts of the economy — a more discriminating pattern of development than we have now.

Reprinted with permission The Toronto Star.

1. In what respect is a "conserver society" similar to the kind of society government officials were hoping for as an answer to the capital shortage (in the preceding section)? In what respect is it different? Are they compatible? Explain.

2. What is the "temple of the gross national product"?

3. Is it true that "economic growth raises the quality of life only to a certain point, and then lowers it"? Why or why not? Give both sides of the argument.

4. Answer Cordell's question about New York City from his point of view and then provide the answer that someone debating him would give, with an explanation and defense of these answers.

5. Provide an economic critique of the logic of Cordell's statement that "the logical thing to do would be to have two model changes per year in the auto industry."

6. What is your feeling about the abandonment of the attempt to compute a "net national product"?

7. What role will be played by the principle of "consumer sovereignty" in the system being proposed in this article? Comment on the value of this principle.

8. Jackson's answer to the question "But aren't such programs only made possible by the surplus wealth which a growth economy produces" is vague and confusing. How would you answer that same question?

9. Explain the logic and implications of Cordell's "upbeat consumer" to "those who see in all this a bleak prospect of reduced living standards and restricted choice."

10. What has the problem of income redistribution to do with the growth in (as opposed to the size of) the national income pie?

11. Although the article mentions the problem of income *re*distribution, it does not address explicitly the problems its suggestions create for income distribution itself. If people no longer earn incomes (by producing junk or pushing paper), how do they buy their groceries? How does this problem fit into their theorizing?

12. According to the last paragraph, zero overall growth is not the target. State in your own words the nature of the growth alternative implied by this article.

II-16 The Future: Theory and Policy

Canada presently faces serious economic problems that could easily become worse in the future. Since traditional policies have proved ineffective, success in solving these problems may well depend on the creation of new institutions capable of attacking these problems in entirely new ways. The first article in this section describes one possible institutional change that may be of value in this respect. The second article pours some cold water on the first article; its contents may be of value in answering some of the questions related to the first article. The third article looks at a new economic problem growing larger and larger on the horizon. We leave to the reader the task of structuring and answering questions related to this article.

II-16 A Corporatism

Toronto *Globe and Mail*, December 10, 1975

Toward a 'new set of values' for Canadians

"It's not the law, it's not the strength of controls which will lick inflation... It's much deeper than that. We want to bring in a new set of values in Canada which will permit Canadians henceforth to develop their great society without resorting to controls."

—Prime Minister Pierre Trudeau speaking to delegates at the Liberal Party convention in Ottawa, Nov. 9, 1975.

BY WAYNE CHEVELDAYOFF
OTTAWA

PRIME MINISTER PIERRE Trudeau has never publicly stated what he thinks the "new set of values" for Canadian society should be and how they will change society.

Until now, Mr. Trudeau has confined himself to explaining that price and income controls are necessary because inflation is being caused by the "excessive" power of the big corporations, big unions and big governments and the exercising of this power must be curbed.

But there is increasing evidence that Mr. Trudeau has a new strategy for changing Canadian society and this is implicitly behind some recent seemingly unrelated federal policy initiatives such as the Canada Labor Relations Council, the call for industry-wide wage bargaining, the controls package and the Royal Commission on Corporate Concentration.

The new goal for society that appears to underlie these initiatives is for labor, management and government leaders to come together at regular intervals to divide up the national income (gross national product) in a sensible, co-operative way that will result in less inflation and fewer disruptive strikes than we now have. The idea has been developed by the in-group of advisers around Mr. Trudeau and its existence inside Parliament Hill's East Block is being perceived by such outside groups as the Canadian Labor Congress.

The new system of tripartite centralized wage bargaining would lead to a drastic transformation of society from the existing system where unions and corporations see themselves as adversaries and the most powerful get the biggest share of national income.

Because the new system would involve a voluntary participation by unions and management, it would be categorized by political scientists as quasi-corporatism, similar to the systems Japan and Sweden already have and toward which West Germany and other European countries are moving. (Pure corporatism is usually defined by political scientists as describing a system where unions and corporations are required by law to participate in national tripartite bodies dominated by the government, as was the case in Italy and Germany during the Hitler and Mussolini eras.)

A related concept is indicative planning, as practiced by France, whereby the government negotiates price and investment targets with industries and selectively controls prices and wages.

Since Mr. Trudeau has never publicly proposed the idea of tripartite co-operation in the dividing up of income, the only evidence that he believes in it is circumstantial. Nevertheless, the evidence is convincing if viewed as a whole.

First, the Prime Minister's own comments suggest he has something more in mind than what he is saying publicly. If the excessive power of corporations, unions and governments is the problem, how will inflation be constrained after controls are lifted in three years or less? The logical implication is that controls will be permanent, or some fundamental change will be made in the economy to change the way power is exercised. In relation to this, it is important to note that the Cabinet's controls package is identical to the one being advocated by Harvard economist John K. Galbraith with one exception: Mr. Galbraith says the controls must be seen to be permanent, while Mr. Trudeau offers them as a temporary solution, possibly hoping this will give them a better chance of being supported by the provinces, labor and business.

Clearly his talk of "a new set of values" is not just a reference to a need to lower inflationary expectations, a subject not even mentioned in the Nov. 9 speech. Rather, the intention is to bring in fundamental social change, perhaps aimed more at influencing other phenomena, such as labor-management strife, than inflation.

Second, Labor Minister John Munro has openly advocated a quasi-corporatist approach for Canada. In a Sept. 19 speech, he said the current work on industry-wide bargaining and harmonizing wage statistics being done by the labor, management and government representatives on the newly formed Canada Labor Relations Council "could ultimately lead to the establishment of a rational tripartite incomes policy for Canada. After all, if management, labor and government can agree on the dimensions of the economic pie, and then sit down at an industry-wide level and determine how big a piece should go to each of them, it is not a far step removed from the sort of equitable distribution of income that a fair incomes policy is designed to bring about.

"That is admittedly looking into the future. For the moment, we must be preoccupied with the necessarily slow, cumbersome and difficult step-by-step process toward that goal." It is not likely that Mr. Trudeau told Mr. Munro to say this; rather Mr. Munro should be credited with perceiving which way Mr. Trudeau's wind is blowing.

Third, the confidential Privy Council document outlining the Trudeau Cabinet's priorities for the coming three or four years puts a particular emphasis on the goal of "co-determination." The concept is essentially that labor should be given greater participation in economic decision-making—at the plant level, in the corporate board rooms and in national economic policy. If labor leaders formally participate in economic policy decisions, then presumably they cannot oppose the decisions. Indeed, they would be in the position of justifying the decisions to their members. Needless to say, the most important economic policy decision that has to be made is: who gets what of the national income pie?

Fourth, a Nov. 24 speech to the Ottawa Rotary Club by Postmaster-General Bryce Mackasey provides the underlying rationale for the "new set of values" although he does not directly advocate tripartite determination of income shares.

Fifth, a senior official came away from a recent meeting with Mr. Trudeau and his close advisers with the distinct impression that Canada "is headed toward corporatism, or syndicalism as the French call it." Rather than finding Mr. Trudeau and others feeling reluctant about the controls program as being a nasty but necessary intrusion into individual freedoms for a short period, he found "an enthusiasm for the idea that labor, corporations and government sit down and divide up the pie."

Although Mr. Trudeau never said it, the startled officials' conclusion was that either the controls program would be permanent, or the controls program is a temporary measure to buy enough time to establish the institutional framework, now lacking, for quasi-corporatism or indicative planning to work.

Arguing that oppression by the employer is a myth and the union adversary role is outdated, Mr. Mackasey said the controls program "could be a blessing in disguise" for unions because it will "shunt the unions off their collision course with society... The union claim for higher wages is no longer a claim against an employer for a larger share of the profit of the company, it's a claim against the public for a higher place in the pecking order ... It's no longer a management-labor conflict where management loses when labor wins, it's a labor-nation conflict with the public interest at stake ...

"Instead of fighting society for a bigger slice of the pie, and thus preventing the pie from being baked, I think the unions should be fighting for a hand in the baking. They should be fighting the root cause of worker discontent. The primary need of the worker today is job satisfaction, a chance to develop his skills and abilities, his initiative, his individuality. The unions should be fighting for a more democratic workplace."

Sixth, the trend toward tripartite co-operation in dividing up national income is on the upswing in Europe and the theoretical debates there are being followed with interest in the Prime Minister's Office. Britain currently has a social contract whereby labor restrains wage demands in exchange for improved Government policies and selective price controls. Economists are predicting this will lead to a more formal tripartite apparatus by 1980. Sweden has had for several years a tripartite system of dividing up national income to profits and wages each year. There is an increasing amount of tripartism in West Germany.

In Canada, this type of system is appealing to those who want price stability and low unemployment—something that orthodox economic theory of a free market economy has not been able to produce. As one senior Government official put it, "the logic is certainly in favor ... Whether it is socially acceptable or proper is another thing."

The logic goes like this: to achieve low unemployment, Canada must promote economic growth; to get that, relying on business to generate employment, Canada must promote capital investment by business, which in turn requires price stability and enough profit. For this, the acquiescence of labor is needed on a voluntary and long-term basis so that high wage demands and labor strife do not defeat the effort to achieve growth.

From another point of view, Canada needs $500-billion in new capital investment over the next decade. To achieve this, labor will have to consume a lower proportion of national income so that a higher proportion can be channelled into investment via corporate profits. Something has to be done, therefore, to obtain labor's co-operation. At this time, power in the labor movement is so fragmented that no single labor organization, such as the CLC, can enter into tripartite negotiations and realistically expect the membership to go along with the result. Hence, the logic implies a new system is needed.

Seventh, at a recent conference of social scientists in Ottawa, Ian Stewart, assistant secretary of the Cabinet in charge of economic policy, remarked that the imposition of controls meant a dramatic change in the functioning of the Canadian economy —a "watershed." Asked to explain, he said that Canada was either headed back to a free market economy, or "toward corporatism, socialism or some sort of collectivism." He did not say which way, but he left little doubt in the minds of observers that it is impossible to revert back to a free market economy.

The second option—corporatism—was what the 17 prominent economists were implicitly arguing against when they sent a letter to Mr. Trudeau saying that controls will not work by themselves and instead the governments should cut their deficits, slow the growth of money supply and reduce the monopolistic and bureaucratic powers of corporations, unions and governments in the Canadian economy. They were, in effect, arguing for a return to a free market economy.

These bits of circumstantial evidence are supplemented by the perception in senior Government circles that Mr. Trudeau and his most senior adviser, Michael Pitfield, Clerk of the Privy Council and Cabinet Secretary, are whole-hearted believers in the Galbraithian view of society.

Mr. Galbraith's view is that the economy has two sectors—the market sector and the planning sector. In the market sector, made up of small businesses, farmers, corner stores, dry cleaners, etc., the firm responds to price changes and swings in credit conditions, and in the interest of making a profit it resists union demands for wage increases.

In the planning sector, where the big corporations and big unions co-operate, the firm has the power to increase prices and does so to pass on rising wage costs to the consumer, with no loss in profit. As a result, the normal tendency in the planning sector is to give in to big wage demands.

To curb the steadily rising wages and prices in the planning sector, the government must intervene with price and wage controls, which can also be extended to some of the market sector, like construction, where there is industry-wide wage bargaining, Mr. Galbraith writes in a recent book, Economics and the Public Purpose. The market sector for the most part, however, can be left alone because competitive forces can be counted on to keep down prices and wages.

Thus, controls for big companies and big unions, as well as for the construction sector, are necessary for price stability and full employment, the Harvard economist maintains. This is exactly what Mr. Trudeau has imposed in Canada, although he has not yet adopted Mr. Galbraith's view that controls should be permanent because inflation will reappear if they are lifted.

(The Trudeau controls package is significantly different than the one proposed by the federal bureaucracy, which favored a more comprehensive approach preceded by a freeze. Mr. Galbraith says a freeze is not necessary to impose a system of controls.)

Among the European theorists being closely read in the Prime Minister's office these days is West German sociologist Ralf Dahrendorf, current head of the London School of Economics and a former chief of the European Economic Commission in Brussels. In a recent book entitled The New Liberty, Mr. Dahrendorf outlines the major political problem of the future as finding ways of peacefully dividing up the national income pie when that pie will be growing more slowly than it has in the past. The current system breeds unequitable shares for the powerless and rapid inflation. Big labor and big corporations, for all the appearance of conflict between them, are really engaged in a tacit conspiracy to secure higher incomes for the industrial sector at the expense of the rest of society.

How is this situation dealt with politically in a democratic society as opposed to an authoritarian regime?

Mr. Dahrendorf believes that a democratic government stands little chance of success in a straight conflict with the big organizations. Former prime minister Edward Heath of Britain lost in a showdown with coal miners and big corporations can simply withdraw their capital from a country if the government puts pressure on them.

Instead, Mr. Dahrendorf argues that the big organizations must be absorbed directly into the apparatus of government, where their leaders will form a kind of senate or second chamber of parliament (or a tripartite council on economic policy?). There is no escape from their power; the only hope is that by making them more visible and compelling them to explain themselves in public they will grow more inhibited in exercising their power.

All this appears to be the theory behind the seemingly fragmented approach by Mr. Trudeau and his close advisers: the controls, at least in the short run, curb the power of the planning sector to achieve price stability and promote capital investment to reduce unemployment; industry-wide bargaining and the Canada Labor Relations Council are an attempt to set up the framework for centralized wage and profit bargaining in a systematic division of the national income pie in the future; the Royal Commission on Corporate Concentration, which has asked businessmen to tell why bigness is good for Canada, brings the conglomerate managers out into the open to justify their concentrated possession of enormous economic power.

The tendency toward tripartite co-operation, indicative planning or quasi-corporatism is there—no matter what it is called.

Why should it worry us? Essentially because a small powerful in-group around Mr. Trudeau could be pushing Canada down the tripartite path without initiating a discussion as to whether that would be desirable.

Will the loss of individual freedoms necessary to make such a system work be worth the projected improvement in unemployment, inflation and labor-management strife? What is desirable about centralized wage bargaining, which would involve little input by the individual worker or union local except for the occasional election of leaders? Why should a small group of corporate executives, union leaders and politicians—all with suspect motives—be given the unprecedented responsibility of dividing up the national income pie? Are investment decisions better left to corporations, or should they be subject to negotiation with Government bureaucrats unfamiliar with business and the marketplace?

---

1. What answers would you provide for the four questions asked by the author in the last paragraph? Defend your answers.

2. Consolidate this article to about a third its length, including only the economic ideas.

3. Write a brief economic critique of the essay produced in the preceding question.

Corporatism Won't Work

Decontrol without restraint?

If the end of wage and price controls depends on a commitment to restraint by both labor and business, the federal government may wait a long time. The fact is that there is no representative body through which either labor or business can make the kind of commitment the government wants.

Ultimately the government will simply have to judge the signs itself to see if there's enough moderation in wages and prices to warrant lifting the controls. It will have to make up its mind soon because the uncertainty about whether and when they'll be removed is, according to most economists, hurting Canada's chances for an economic recovery.

The economy is running at something like 80 to 85 per cent capacity and industry is not going to commit itself to expansionary policies while it's in the straitjacket of controls.

Prime Minister Trudeau seems to have accepted this thesis for he said this week that decontrol before the anti-inflation program windup date of December 31, 1978, is "possible and perhaps even desirable." But he said he still wanted a commitment to restraint.

The government has indeed tried to tie decontrol to the development of a tripartite approach to economic decision-making in the post-control period, involving labor, business and government in a central planning organization. There has been a series of meetings over the last few months which began optimistically but have since become more and more pessimistic about the prospects of success. Tripartism, it seems, was a good idea that fell apart when the participants began to explore it in detail.

It become clear, for instance, that the Canadian Labor Congress had no constitutional authority to commit its affiliated unions to any policy of restraint since the unions are autonomous in all matters of collective bargaining. Furthermore, there was no consensus among the unions that labor's participation would be acceptable to them all.

The unions are afraid that their freedom to bargain contracts would be compromised by their participation in such a planning organization and that, in fact, collective bargaining might to all intents and purposes disappear.

Similarly business feels that all its disparate interests can't be adequately canvassed by a few representatives in the tripartite group. And if the group were widened to meet those objectives it would become too unwieldy to be effective.

The government, too, is having second thoughts about a group having authority to make national economic decisions that ought to be made by Parliament where the interests of the public as well as those of labor and business are asserted. In these circumstances a tripartite organization would be nothing more than an advisory group and there are already enough of them.

Restraint was to have been the response of labor and business to the offer of a share in joint economic decision-making. Since a tripartite organization is not on the cards, the government can't expect a "deal" from them. That, together with the problems in seeking commitments from representatives who can't make them, make it unreasonable for the government to expect commitments to restraint. It will simply have to make up its own mind whether to start the decontrol process now or leave the controls on until the end of 1978.

C Separatists' Economics

Financial Times of Canada, August 14, 1977

Federalist weapon may just boomerang

By JOAN FRASER
Financial Times

MONTREAL — Most English Canadians still do not realize it, but economics could turn out to be an enormous boomerang for federalists in the struggle for national unity.

In English Canada, the conventional wisdom still holds that federalists have all the good economic arguments and when the crunch comes, that will be one of the major reasons why Quebec will decide to stay in Canada.

But in Quebec many economists have concluded that almost all of Ottawa's economic policies since Confederation have actually hurt Quebec's development.

More important, they believe that unless profound changes occur, federal policies will continue to stunt the province's economic growth.

Pierre Fortin, an economics professor at Laval University, goes so far as to say that if the rest of Canada does not wake up to this Quebec sentiment, economic grievances could tip the balance against Confederation in the coming referendum on Quebec's independence.

And Gilles Paquet, an economics professor at Carleton University, warns that unless the system changes, Quebec may have more to gain than to lose by becoming independent.

These arguments are difficult for English Canadian federalists to swallow, in part because the Quebec economists are looking only at Quebec's point of view.

The core of the Quebec argument is put by Pierre Frechette, an economics professor at the University of Montreal: "Every sectorial policy adopted since the National Policy (in 1889) has had multiple (adverse) consequences, notably on the profitability of new industries in Quebec."

Yves Rabeau, vice-chairman of the university's centre for research on economic development and one of the most influential members of this school of thought, says "it is not a question of policies systematically designed to hurt Quebec.

"It is a question of bad planning and a lack of imagination. Policies are brought in without being thought through, and then they turn out to be harmful." The great flaw, Rabeau thinks, has been Ottawa's failure to provide for regional "equalization of economic benefits" of its policies

By this analysis, better federal policies would have wiped out the need to give Quebec the massive revenue equalization payments often cited as evidence of the provinces's net gain from Confederation, because Quebec would now have a healthy economy instead of one in decline.

The summary which follows draws together work by a wide range of economists; most agree with the broad lines of the general analysis.

The view it presents is at best unpalatable and at worst infuriating for federalists in the rest of the country. But it offers powerful ammunition for separatists.

The National Policy raised tariff barriers to foster native industry. This is generally thought to have benefited Quebec and Ontario at the expense of the rest of the country.

But Albert Faucher, a respected economic historian at Laval, has argued that because Ontario had access to iron and coal resources, plus the booming western and mid-west markets, that was where capital-intensive heavy industry located, laying the foundation for Ontario's present wealth.

What Quebec got was industries based on resource exploitation plus a series of small, inefficient manufacturing plants relying on the province's large pool of cheap labor.

An example of the latter is the textile industry, which needs ever-increasing protection and which still employs 118,000 Quebeckers. That kind of native industry, the economists think, is the kind Quebec could have done without.

And apart from building tariff walls, Ottawa did little to ensure that industrial development spread evenly through the country instead of concentrating in Ontario. Thus, only 30 per cent of Quebec's exports now are finished products, compared to 69 per cent for Ontario.

Quebec historically had one giant natural advantage: Ocean ships could not sail past Montreal, which thus became Canada's major port of entry. The St. Lawrence Seaway destroyed that advantage.

Rabeau has shown, using Stat-Can figures, that completion of the Seaway in 1958 hurt Montreal badly. In 1959, new industrial investments in Toronto were only 55 per cent of the Montreal total; in 1966, investments in Toronto equalled 165 per cent of the Montreal total. Per capita manufacturing investment in Montreal has never regained the peak levels of 1959.

Until 1961 Montreal had a flourishing refinery industry based on imported oil and supplying much of eastern Canada. But in that year, after the report of the Borden commission (to which Quebec was not invited to send a delegate), Ottawa gave Alberta gas and oil a monopoly on all markets west of the Ottawa river.

While Sarnia's refineries blossomed, Montreal's stagnated.

Ontario was the big beneficiary of the auto pact. Only General Motors of Canada Ltd. built an assembly plant in Quebec, and it was decided on before the pact was signed.

Rabeau says Ottawa should have required that all expansion of the Canadian auto industry be diversified among the provinces, with Quebec getting 25 per cent or 30 per cent of new plants.

Federal policy has fostered the growth of a highly centralized (and now Toronto-based) financial system. Andre Ryba, an economist with the Bank of Montreal, has found that "Quebec financial markets have ... become extensions of the Ontario market," a development favored by the "neutral" policy of the Bank of Canada.

Several economists, notably Francois Dagenais (now a provincial civil servant), think federal freight rate and other policies on feed grains have prevented Quebec from developing a native beef industry. They say that if all subsidies were eliminated and market forces ran free, Quebec could be self-sufficient in beef.

Federal budgetary policy is mainly geared to the "big numbers" — the national growth, unemployment and inflation rates. But national numbers disguise the fact that when the Ontario economy is over-heating, the eastern provinces often are still deep in recession.

The Department of Regional Economic Expansion was designed to ease regional disparities. But its subsidies often bear little relation to the need to establish a sound industrial base.

A 1973 study showed that only 24 per cent of DREE grants in Quebec went to investments which met all provincial criteria for economic growth, and 29 per cent went to investments which met none.

Not until this summer did DREE launch a program to attract high-potential industries to Montreal, Quebec's natural growth centre — and that program, too, is temporary.

Gilles Paquet thinks one basic problem is the federal bureaucracy, which he notes has assumed "incredible power" through such bodies as the Foreign Investment Review Agency and the Anti-Inflation Board.

These civil servants are the only people who really understand — and hence control — the details of federal-provincial battles. And like bureaucrats anywhere (including those in Quebec City), they will resist any erosion of their empires with all the strength they can muster.

Glossary

Ad Valorem Tax – a tax on a good or service which is a fixed percentage of the value of the good or service, e.g., a sales tax.

Arc Elasticity – elasticity calculated using averages of the before and after figures as bases for calculation of percentage changes.

Automatic Stabilizer – nondiscretionary or "built-in" features of government revenues and expenditures that automatically cushion recession by helping to create a budget deficit and curb inflation by helping to create a budget surplus, e.g., unemployment insurance payments and benefits, income tax, and agricultural price supports.

Average Propensity to Consume (APC) – the ratio of consumption to income.

Average Total Cost (ATC) – total cost divided by the number of units of output.

Average Variable Cost (AVC) – total variable cost divided by the number of units of output.

Balance of Payments – the difference between a country's annual supply of foreign exchange and its annual demand for foreign exchange.

Balance of Trade – the difference between a country's exports and imports (of goods).

Balanced-Budget Multiplier – the multiplier associated with an increase (or decrease) in government spending financed by an equal increase (decrease) in taxes.

Bank of Canada – Canada's central bank.

Bank Rate – the interest rate at which the Bank of Canada will loan cash reserves to the chartered banks. The comparable U.S. rate is called the *discount rate*.

Bellwether – a signalling device.

Bottleneck Inflation – *see* Demand shift inflation.

Bretton Woods – site of a 1944 international monetary conference at which the postwar world fixed exchange rate system was structured and the International Monetary Fund (IMF) was created.

Broad Definition of Money – *see* Money.

Capital – a factor of production, defined to include all man-made aids to production such as buildings, equipment, and inventories but not stocks, bonds, mortgages or money.

Capital Account – the difference between a country's annual capital inflows and capital outflows.

Capital Inflows – a supply of foreign exchange to a country arising from foreigners purchasing that country's financial assets or businesses. This would be a *capital outflow* to the foreign countries in question.

Capital Stock – the aggregate quantity of an economy's capital goods.

Capitalize – the calculation of a lump sum equivalent of a series of payments.

Cartel – an organization of producers designed to limit or eliminate competition among its members; the term generally applies to government-enforced agreements.

Cash Reserves – *see* Reserves.

Central Bank – a bank that acts as banker to the banking system and to the government. It is the sole money-issuing authority.

Ceteris Paribus – literally "other things being equal". Used to indicate that all variables except those specified are assumed unchanged.

Chartered Bank – a bank licensed as such by Parliament under the Bank Act; they are subject to considerable regulation by the Bank of Canada. There are only about a dozen chartered banks in Canada. The comparable U.S. term is *commercial bank* (of which there are thousands).

Cobweb – *see* Corn—hog cycle.

Commercial Paper – short-term debt issued by well-known corporations.

Complement – goods used in conjunction with one another are complementary goods (ham and eggs); if the price of one good rises (falls) the demand for the other good falls (rises).

Consumers Surplus – the difference between the amount paid for a commodity and the highest amount the consumer would be willing to pay to avoid going without.

Corn—Hog Cycle – a repetition of the sequence of low output and high prices followed by high output and low prices. These cyclical fluctuations in price and quantities arise for certain agricultural products where the quantity supplied to the market is determined at a previous time when production plans have to be formulated.

Corporatism – a system in which labor, business and government leaders meet regularly and negotiate the division of national income.

Cost—Benefit Analysis – an attempt to estimate the total costs and total benefits to society of public programs or policies.

Cost—Push – a theory of inflation that rests on cost increases caused by factor payments increasing faster than productivity or efficiency: This inflation is unrelated to excess aggregate demand.

Crowding-Out – decreases in private aggregate demand associated with the means used to finance an expansionary fiscal policy.

CSB – Canada savings bond.

Current Account – the balance of trade augmented by the net foreign exchange transactions associated with services.

Demand Deposit – a bank deposit that is withdrawable on demand and transferable by means of a cheque.

Demand—Pull – a theory of inflation that rests on excess aggregate demand.

Demand—Shift Inflation – a theory of inflation that rests on a lack of matching of the components of aggregate demand and aggregate supply.

Depreciation – the decline in the value of capital due to its wearing out or becoming obsolete.

Depression – a prolonged period of very low economic activity which is generally characterized by large-scale unemployment of resources.

Devaluation – a drop in the value of a country's currency on international exchange markets (or a rise in the cost of buying foreign exchange).

Dirty Floater – a country whose government buys or sells foreign exchange so as to influence the market-determined value of its currency.

Dumping – sale of the same product in different markets at prices lower than in the "home" market; generally associated with selling a good in a foreign country at a price below the domestic price.

Dynamic – relating to movement over time. Rather than simply comparing old and new equilibrium positions (comparative statics), a dynamic analysis examines the character of the economy's movement from the old to the new position.

Economies of Scale – a decrease in a firm's long-run average costs as the size of its plant is increased.

Efficiency – the degree to which the economy achieves a position in which it is impossible to reallocate or redistribute goods and services so as to make someone better off without making someone else worse off.

Elasticity – see Price elasticity.

Envelope – a curve formed by joining the extremities of a family of curves.

Equilibrium – a state of balance in which there are no internal pressures for change, usually characterized by equality between demand and supply.

Excess Demand – demand exceeding supply at a given price. The reverse is called excess supply.

Exchange Rate – the price of one currency in terms of another.

Exogenous – autonomous, not originating from within the system.

External Diseconomies – by-products (costs) of a firm's production process for which it does not pay.

External Economies – by-products (benefits) of a firm's production process for which the firm is unable to charge.

Federal Reserve Board – the central bank of the United States.

Fiscal Policy – a change in government spending or taxing, designed to influence economic activity.

Fixed Costs – costs that do not change as the firm changes its level of output.

Fixed Exchange Rate – an exchange rate held at a constant level by a government promise to buy or sell foreign exchange at that rate.

Flexible Exchange Rate – an exchange rate whose level is determined by the unhindered market forces of supply and demand for foreign exchange.

Floating Exchange Rate – see Flexible exchange rate.

Foreign Exchange Fund – a fund containing the government's holdings of foreign currency or claims thereon.

Forward Exchange Market – a market in which foreign exchange can be bought or sold for delivery (and payment) at some specified future date but at a price agreed upon in the present.

Frictional Unemployment – unemployment associated with people changing jobs or quitting to search for a new job.

Futures Market – a market in which a good can be bought or sold for delivery (and payment) at some specified future date but at a price agreed upon in the present.

GNP – gross national product, the total of all final goods and services produced during the year.

Hedging – a technique of buying and selling future contracts that minimizes the risk of loss due to price fluctuations.

Hog Cycle – see Corn-hog cycle.

Indifference Curve – a graph of combinations of quantities of two goods, with any of which a consumer is equally satisfied; each of these combinations yields the same total utility to the recipient.

Labor Force – the number of people either employed or actively seeking work.

Liquidity – the ability to meet current financial liabilities in cash.

Liquidity Preference – a theory of the determination of the interest rate, resting on the supply of and demand for money.

Loanable Funds – a theory of the determination of the interest rate, resting on the supply of and demand for bonds and other financial assets.

M1, M2 – see Money.

Marginal Benefit – the extra benefit created by increasing or decreasing output (consumption) by one unit.

Marginal Cost – the extra cost resulting from increasing output by one unit.

Marginal Propensity to Consume – the proportion of an extra dollar of income that is spent on consumption goods and services.

Marginal Revenue – the change in a firm's revenue resulting from selling one additional unit of output.

Marginal Utility – the additional satisfaction gained (or lost) by a buyer from consuming one unit more of a good.

Marketing Board – an association of producers set up, under government sanction, to market output jointly.

Mixed Economy – an economy in which some decisions as to what to produce, and for whom to produce are made by firms and households and some by central authorities.

Monetarists – economists who believe that fluctuations in the quantity of money are the primary causes of economic fluctuations.

Monetary Policy – a change in a monetary variable, such as the quantity of money, designed to affect economic activity.

Money – the narrow definition of the money supply (M1) is the sum of currency and demand deposits. The broad definition (M2) is M1 plus time deposits. *See also* Printing money.

Money Market – the market in which money is borrowed and loaned; the stock and bond market.

Money Rate of Interest – the observed interest rate in the money market.

Monopolistic Competition – the structure of an industry with many sellers and freedom of entry, each seller having a product slightly different from the other, giving him some control over his price.

Monopoly – an industry characterized by a single seller.

MPC – see Marginal propensity to consume.

Multiplier – the amount by which national income increases when government spending increases by one dollar. In more general terms, the factor by which the magnitude of a policy action must be multiplied to give the impact of that policy on some specified dimension of economic activity.

Narrow Definition of Money – see Money.

Natural Rate of Unemployment – the unemployment rate to which it is thought the economy will gravitate over the long run. Its magnitude is determined by institutional factors, such as the degree to which changes in tastes and technology maintain structural unemployment, the degree to which information problems and geographical immobility maintain frictional unemployment, and the levels of the minimum wage and UIC benefits.

Nominal Rate of Interest – see Money rate of interest.

Oligopoly – a market structure in which a small number of rival firms dominate the industry.

OPEC – Organization of Petroleum Exporting Countries.

Open Economy – an economy that engages in foreign trade.

Open-Market Operations – the purchase or sale of securities on the money market by the central bank with the goal of expanding or contracting the supply of money and credit.

Opportunity Cost – the return a resource could earn in its most profitable alternative use.

Participation Rate – the percentage of the noninstitutionalized population over age 15 that is in the labor force.

Peripheral Work Force – workers who are not the primary breadwinners in their household.

Phillips Curve – a relation between the rate of change of money wages and unemployment, often drawn as a relation between inflation and unemployment.

Price Discrimination – the sale by a firm of the same commodity to different buyers at different prices, for reasons unrelated to cost.

Price Elasticity of Demand – the percentage change in quantity demanded caused by a price change (*ceteris paribus*), divided by the percentage change in price, conventionally expressed as a positive number.

Price System – the use of freely-determined market prices to direct the allocation and distribution of goods and services within an economy.

Prime Rate – the interest rate charged by chartered banks to their most favored customers.

Printing Money – the sale of bonds by the government to the central bank.

Producers' Surplus – revenue received by producers, in excess of their actual production cost.

Progressive Tax – a tax which, as a percentage of income, falls more heavily on those with higher incomes.

Quota System – a production system in which each producer is permitted to produce only a specified quantity, called his quota.

Real Income – money income corrected for changes in the price level (the purchasing power of money income).

Real Rate of Interest – the money rate of interest less the expected rate inflation (the money rate of interest corrected for the expected change in the purchasing power of money).

Real Wage – the money wage corrected for price level changes.

Recession – a downswing in the level of economic activity.

Reflation – the process of managing money for the purpose of restoring a previous price level.

Rent – the return to a factor in excess of the return required to entice that factor to its present employment.

Reserves – in banking, that part of customers' deposits kept by a bank either as currency held in its own vault or as deposits with the central bank.

Sales Tax – a tax levied as a percentage of retail sales.

SDR – special drawing right, the name given to the "currency" of the International Monetary Fund.

Spot – for immediate payment and delivery, as opposed to future payment and delivery.

Stagflation – the simultaneous existence of high inflation and high unemployment.

Sterilization – central bank action offsetting money supply changes brought about, under a fixed exchange rate system, by balance of payments deficits or surpluses.

Structural Unemployment – unemployment resulting from technological displacement, lack of appropriate skills, and geographic imbalances between the supply of and demand for labor.

Substitute – two goods are said to be substitutes if an increase (decrease) in the price of one leads to a rise (fall) in the demand for the other.

Swap Agreements – holdings of U.S. Treasury securities by the Bank of Canada, purchased by the Bank from the government's Exchange Fund Account, with a promise by the government to buy these securities back at a set price.

Tariff – a tax applied on import which can be based on a tax per unit of the commodity or on the value of the commodity.

Term Deposit – a savings deposit where the depositor specifies the length of time the money is to be deposited. It generally can be withdrawn before its maturity date with a penalty such as a lower interest rate paid. Many writers use term, time and notice deposits interchangeably.

Time Deposit – an interest-earning bank deposit, subject to notice before withdrawal; also called a notice deposit.

Treasury Bill – the characteristic form of short-term government debt.

Tripartism – *see* Corporatism.

UIC – Unemployment Insurance Commission.

Utility – consumer satisfaction.

Variable Costs – costs whose total varies directly with the level of output.

Velocity – the ratio of GNP to the money supply.